Timber Design

Timber Design covers timber fundamentals for students and professional architects and engineers, such as tension elements, flexural elements, shear and torsion, compression elements, connections, and lateral design. As part of the *Architect's Guidebooks to Structures* series, it provides a comprehensive overview using both imperial and metric units of measurement. *Timber Design* begins with an intriguing case study and uses a range of examples and visual aids, including more than 200 figures, to illustrate key concepts. As a compact summary of fundamental ideas, it is ideal for anyone needing a quick guide to timber design.

Paul W. McMullin is an educator, structural engineer, and photographer. He holds degrees in mechanical and civil engineering and is a licensed engineer in numerous states. He is a founding partner of Ingenium Design, providing innovative solutions to industrial and manufacturing facilities. Currently an adjunct professor at the University of Utah in Salt Lake City, USA, he has taught for a decade and loves bringing project-based learning to the classroom.

Jonathan S. Price is a structural engineer and adjunct professor at Philadelphia University in Pennsylvania, USA, where he was honored with the Distinguished Adjunct Faculty Award in 2006. He holds a bachelor of architectural engineering degree from the University of Colorado, USA, and a master of science degree in civil engineering from Drexel University in Philadelphia, USA, and is registered in 12 states.

Architect's Guidebooks to Structures

The *Architect's Guidebooks to Structures* series addresses key concepts in structures to help you understand and incorporate structural elements into your work. The series covers a wide range of principles, beginning with a detailed overview of structural systems, material selection, and processes in Introduction to Structures, following with topics such as Concrete Design, Special Structures Topics, Masonry Design, and Timber Design, and finishing with Steel Design, to equip you with the basics to design key elements with these materials and present you with information on geotechnical considerations, retrofit, blast, cladding design, vibration, and sustainability.

Designed as quick reference materials, the *Architect's Guidebooks to Structures* titles will provide architecture students and professionals with the key knowledge necessary to understand and design structures. Each book includes imperial and metric units, rules of thumb, clear design examples, worked problems, discussions on the practical aspects of designs, and preliminary member selection tables, all in a handy, portable size.

Read more in the series blog: http://architectsguidestructures. wordpress.com/

Introduction to Structures
Paul W. McMullin and Jonathan S. Price

Concrete Design
Paul W. McMullin, Jonathan S. Price, and Esra Hasanbas Persellin

Special Structural Topics
Paul W. McMullin and Jonathan S. Price

Masonry Design
Paul W. McMullin and Jonathan S. Price

Timber Design
Paul W. McMullin and Jonathan S. Price

Steel Design
Paul W. McMullin, Jonathan S. Price, and Richard T. Seelos

Timber
Design

**Edited by Paul W. McMullin
and Jonathan S. Price**

NEW YORK AND LONDON

First published 2017
by Routledge
711 Third Avenue, New York, NY 10017

and by Routledge
2 Park Square, Milton Park, Abingdon, Oxon OX14 4RN

Routledge is an imprint of the Taylor & Francis Group, an informa business

Library of Congress Cataloguing in Publication Data
Names: McMullin, Paul W., author. | Price, Jonathan S., author.
Title: Timber design / Paul W. McMullin and Jonathan S. Price.
Description: New York : Routledge, 2017. | Includes bibliographical
references and index.
Identifiers: LCCN 2016038254| ISBN 9781138838703 (hardcover) |
ISBN 9781138838710 (pbk.)
Subjects: LCSH: Building, Wooden. | Structural design.
Classification: LCC TA666 .M39 2017 | DDC 691/.1—dc23
LC record available at https://lccn.loc.gov/2016038254

ISBN: 978-1-138-83870-3 (hbk)
ISBN: 978-1-138-83871-0 (pbk)
ISBN: 978-1-315-73389-0 (ebk)

Acquisition Editor: Wendy Fuller
Editorial Assistant: Norah Hatch
Production Editor: Alanna Donaldson

Typeset in Calvert
by Florence Production Ltd, Stoodleigh, Devon, UK

For our mentors
John Masek, Steve Judd, Terry Swope,
Suzanne Pentz, Marvin F. Ricklefs,
Robert Brungraber

"*Timber Design* is a down-to-earth guide for students and practitioners alike. It covers the range of contemporary timber materials and provides in-depth descriptions of design methodologies in all modes of application. A most useful book, it deftly combines theory and practice."

Brian J. Billings, RA, Adjunct Professor of Architecture, Philadelphia University, USA

Contents

Contributors

Sarah Simchuk is a project architect and fine artist working towards architectural licensure in large-scale retail design. She holds bachelor's and master's degrees in architecture from the University of Utah. She is in the early stages of her architectural career, with an inclination towards design and details in project management. She comes from a fine art background, with more than 15 years' experience in hand drawing and rendering, and lends a 3-D approach to the understanding of structures.

Hannah Vaughn is an architect based in Salt Lake City, UT. She is a dedicated practitioner and teaches as adjunct faculty at the University of Utah School of Architecture.

Frank P. Potter, P.E., S.E., LEED AP, has been a practicing structural engineer for more than 20 years. He holds a bachelor of science degree in civil engineering from the University of Wyoming and is licensed in several western states. He has been a principal in a structural engineering firm in Salt Lake City. Currently, Mr. Potter is the western region engineer for Boise Cascade and lectures on the benefits and applications of engineered wood. He enjoys living in the west, spending much of his spare time fly fishing, skiing, and mountain biking.

Acknowledgments

Like previous books in the series, this volume wouldn't be what it is without the diligent contributions of many people. We thank Sarah Simchuk and Hannah Vaughn for their wonderful figures and diligent efforts; Frank Potter and Thomas Lane for reviewing each chapter; and Esra Hasanbas for her help with the examples.

We thank Wendy Fuller, our commissioning editor, Norah Hatch, our editorial assistant, Louise Smith, our copy editor, Alanna Donaldson, our production editor and Julia Hanney, our project manager. Each of you have been wonderful to work with and encouraging and helpful along the journey. We thank everyone at Routledge and Florence Production who produced and marketed the book.

Special thanks to our families, and those who rely on us, for being patient when we weren't around.

We are unable to fully express our gratitude to each person involved in preparing this book. It is orders of magnitude better than it would have otherwise been, thanks to their contributions.

Introduction

The human race has used timber for centuries to provide simple shelter and proud monuments. It's lightweight, soft nature makes it workable by hand—ideal when large machines are unavailable. We find timber in millions of homes around the world, in smaller commercial structures, and frequently in the great structures of history. Properly cared for, it can last for centuries.

Today, engineered wood products have greatly expanded the sustainability and versatility of timber structures. By placing stronger material—and more of it—in the right places, we are able to take greater advantage of wood's best properties. This allows us to span the same or greater distances as solid sawn timber, but using smaller trees.

This guide is designed to give the student and budding architect a foundation for successfully understanding and incorporating timber in their designs. It builds on *Introduction to Structures* in this series, presenting the essence of what structural engineers use most for timber design.

If you are looking for the latest timber trends, or to plumb the depths of technology, you're in the wrong place. If you want a book devoid of equations and legitimate engineering principles, return this book immediately and invest your money elsewhere. However, if you want a book that holds architects and engineers as intellectual equals, opening the door of timber design, you are very much in the right place.

Yes, this book has equations. They are the language of engineering. They provide a picture of how structure changes when a variable is modified. To disregard equations is like dancing with our feet tied together.

This book is full of in-depth design examples, written the way practicing engineers design. These can be built upon by examples being reworked in class, with different variables. Better yet, assign small groups of students to rework the example, each with new variables. Afterward, have them present their results and discuss the trends and differences.

For learning assessment, consider assigning a design project. Students can use a past studio project, or a building that interests them. The project can start with students determining structural loads, continue with them designing key members, and end with consideration of connection and seismic design. They can submit design calculations and sketches summarizing their work and present their designs to the class. This approach requires a basic level of performance, while allowing students to dig deeper into areas of interest. Most importantly, it places calculations in context, providing an opportunity for students to wrestle with the iterative nature of design and experience the discomfort of learning a new language.

Our great desire is to bridge the gap between structural engineering and architecture, a gap that historically didn't exist, and is unnecessarily wide today. This book is authored by practicing engineers, who understand the technical nuances and the big picture of how a timber project goes together. We hope it opens the door for you.

Merion Friends Meetinghouse

Chapter 1

Jonathan S. Price

1.1 INTRODUCTION

Merion Friends Meetinghouse, shown in Figure 1.1, is located in southeastern Pennsylvania and is the oldest Quaker meetinghouse in the state. It is the second oldest in the U.S., eclipsed in age by the Third Haven Meetinghouse in Easton, Maryland (c. 1682). It was a religious and community center for centuries and is still used today.

1.2 HISTORICAL OVERVIEW

The Merion settlement was on land given to young William Penn by King Charles II. Penn received this land as repayment of a loan Penn's father had given to the king in 1660 so that England could rebuild its navy. This was a win–win for the king, who now could get rid of the troublesome Quakers and pay off a substantial debt. For William Penn, this was an opportunity to realize his dream of planting "the seed of a nation," reflecting Quaker ideals in the New World.[1] In 1682, the first group of

Figure 1.1 Merion Friends Meetinghouse, front entrance
Source: Photo courtesy of Keast & Hood

Jonathan S. Price

Quakers from Merionethshire, Wales, settled near Philadelphia and, in 1695, they began construction of the meetinghouse.[2] Although the land was given to Penn by the king, he paid the Native Americans £1,200 for it, rather than take it through conquest.[3]

1.3 BUILDING DESCRIPTION

The meetinghouse is a modest structure and is nontraditional because of its T-shaped plan, as most meetinghouses are rectangular. Because Merion has two ridgelines meeting at a rather central point, there are valleys that allow accumulations of leaf debris, making the roof susceptible to leaks.

The roof **frames** are Welsh-influenced **cruck**-type frames supported on 24-in-thick stone walls, and these frames resemble A-frames. They have **truss elements**, which were intended to **support** the original high ceiling and perhaps restrain outward thrust. The low ceiling was added in 1829 (ref. Figure 1.2 and longitudinal section of Figure 1.3).

Some believe that the southern portion (stem of the tee) was constructed first, with the northern section an addition. No evidence has been found to support this theory, such as remnants of an old foundation in the crawl space. Also, the north section center frame was tenoned into the first frame of the south section, and so it must have followed soon after the south (Figure 1.4).[4]

Before the meetinghouse was nominated to the National Historic Register, it attracted attention. In 1981, noted historian David Yeomans wrote:

> The most interesting roof to have been found in this area [i.e., Pennsylvania, New Jersey, and Delaware] is that of Merion Meetinghouse because this structure uses a primitive form of trussing I have not so far seen in England, although it clearly derives from there. The principal rafters curve downward sharply at their feet—a feature shown in the earliest published drawing of a roof structure. The tie beams are trussed up with timbers that are not quite **king posts** in that they are not hung from the apex of the roof. No metal strapping is used and instead the post is fixed to the tie beam with a dovetail.[5]

Ideally, the original curved frame pieces would have been sawn from trees with large sweeping branches, so that the **grain** and **stresses** could have followed the curve. Instead, they used large sections of straight

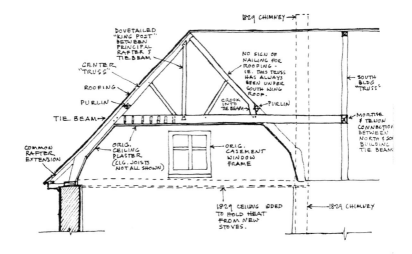

SECTION A-A - SHOWING CENTER "TRUSS" TIE BEAM
CONNECTING NORTH & SOUTH BUILDINGS.

MERION FRIENDS MEETING HOUSE, MERION, PA.

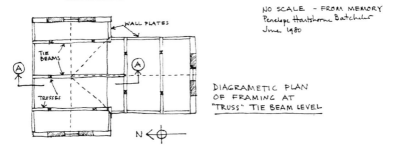

NO SCALE - FROM MEMORY
Penelope Hartshorne Batcheler
June 1980

DIAGRAMETIC PLAN
OF FRAMING AT
"TRUSS" TIE BEAM LEVEL

Figure 1.2 Unpublished survey drawings by Penny Batcheler, c. 1980

Jonathan S. Price

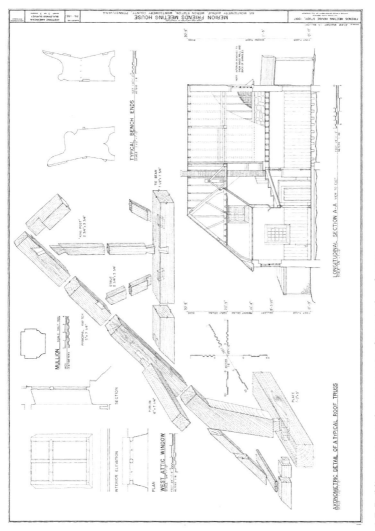

Figure 1.3 Building section and axonometric of a typical roof truss

Source: Historic American Building Survey, c. 1997

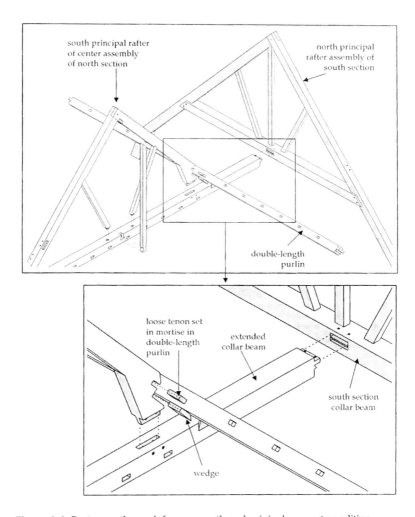

Figure 1.4 Center north cruck frame—south end original support condition

Source: David Mark Facenda, "Merion Friends Meetinghouse: Documentation and Site Analysis," Thesis for Master of Science in Historic Preservation, University of Pennsylvania, 2002, p. 114

Jonathan S. Price

timber. Perhaps this was to avoid breaking an English naval ordinance dated April 22, 1616:

> Crooks, Knees, and Compass timber . . . will be of singular Use for the Navy, whereof principal Care is to be had, in order to the Kingdom's Safety: It is therefore Ordered and Ordained, by the Lords and Commons in Parliament assembled, That the Crooks, Knees, and Compass Timber, arising from any Trees felled for any of the said Services by Order from the Committee of His Majesty's Revenue, be reserved to the Use of the Navy, and not disposed of to any other Use.[6]

Compression and **shear forces** in a curved timber cause it to bend. If the grain does not follow the curve, then **tension** stresses will develop across the grain (see Figure 1.3). Factors of safety built into modern **codes** allow for some nonparallel grain at **knots** but not across the entire cross section, which woud result in a significant strength reduction.

1.4 SURVEY AND ASSESSMENT

John Milner Architects Inc., of Chadds Ford, PA (Daniel Campbell, AIA), retained Suzanne Pentz, the director of historic preservation with Keast & Hood engineers, to assess the building structure following an observation made by a roofer regarding the north wall curvature. The roofer asked if they were to follow the curvature of the supporting wall and the roof edge or to lay the shingles in parallel i.e. straight lines.

During the structural investigation and assessment, Unkefer Brothers Construction Company helped by removing wall finishes where observations were required (i.e., probes).

We discovered that the base of several cruck frames and sill plates that were coincident with the roof valleys had decayed. The north wall was leaning—out of plumb by about one-third of its thickness. Other discoveries included ineffective framing modifications made in the 1800s that will be further discussed, loose king posts, plus rot and termite damage within the crawl space. To quantify the amount of decay at frame bases and wall plates, the assessment included resistance drilling, which allows the investigator to look for decay within the timber.

1.5 ANALYSIS

The north wall displacement was sufficient evidence that the cruck frame supports were yielding, caused by an outward thrust. The plumbness and

Resistance drilling is based on the principle that biological decay of wood is consistently accompanied by reduction in density and therefore in resistance to mechanical penetration. Although the technique itself has been known for decades, it has been recently facilitated by the introduction of the 'Resistograph', a proprietary instrument made by Instrument Mechanik Labor (IML) of Wiesloch, Germany. [7]

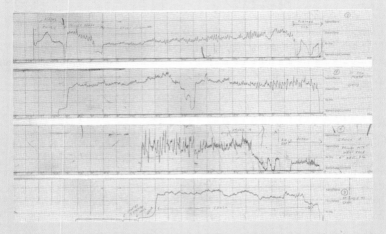

Figure 1.5 Resistograph drill logs. High points are where material is dense; low points indicate decay or voids

Source: Image courtesy Keast & Hood

bowing measurements of the north wall and computer modeling of the entire structure confirmed our assumptions that the structure was in trouble.

The computer models also confirmed that support assumptions were dramatically influencing the results. We first modeled a typical frame using a 2-D approach, with one support **pinned** and the other on a **roller**, but this predicted incipient failure. We knew the walls were not entirely **rigid** (**fixed**), and so we iterated a more detailed 3-D computer model of the entire structure (frames and purlins), assuming varying degrees of lateral **stiffness** of the walls, until the displacements agreed with field

Jonathan S. Price

measurements. The predicted **lateral** thrust exceeded the wall's resistance, and therefore another mechanism was assisting.

It may come as a surprise to the reader that structural characteristics of supports are often based on assumptions, which are, in turn, based on experience. Supports that are neither rigid nor totally yielding are difficult to quantify.

We concluded that there were **redundant load paths** in the roof timbers and **diaphragm**, or else the frames would have failed. In our report, we acknowledged that we could not exactly determine how **loads** were resisted. We were humbled by the unknowns and observed that:

> **Analysis** is an exercise in evaluating the evidence. Confidence in the results is proportional to the simplicity of the framing. More complicated and redundant **systems**, such as the truncated center cruck frame, resist direct analysis. By necessity, an iterative approach was needed to explain the observations and to 'bracket' the results. Unfortunately, no amount of computer horsepower can assuage concerns with the reduced southerly support of this frame.[8]

1.6 STRUCTURAL STRENGTHENING

Without 100 percent confidence in our analysis, we moved forward with repair and preservation plans. The main concern was how to address the well deflections plus repair of deteriorated fabric. Multiple preservation objectives were on the table, not just structural, and so the Friends asked us to prioritize: first on the list was stability of the north wall and roof framing, followed by the ground floor deterioration. Unkefer Brothers' estimates confirmed that funding for only the most pressing issues was available.

The initial plans included local reinforcement and replacement of deteriorated material with new material (Dutchman repairs), new tie-rods to contain outward thrust, and reinforcement of the framing around the chimney with steel channels. These repairs addressed the various issues but would have required extensive shoring of all affected roof framing, from the ground floor up (and through the crawl space to the soil below). The plans and details were finalized, and construction estimates were prepared.

After this event, another approach occurred to the design team. Rather than addressing the repairs individually, we coalesced them into one fundamental approach.

1.7 A NEW PLAN

Instead of treating the seemingly disparate problem areas—that is, decaying cruck frame bases, unfortunate chimney work-around framing—we offered a new solution—a truss under the north ridge line lifting the three cruck frames and thereby relieving stress on all of them plus the chimney work-around (shown in Figure 1.6). This truss was to be assembled by hand in the attic, but had the added advantage of eliminating the shoring, and created a net savings of $37,000.

Replacement of decayed wall plates was still included, but without the need for shoring, and the solution was reversible. We quickly revised the documents and moved forward with the new plan.

1.8 LESSONS LEARNED

Several lessons stand out from this experience.

1. When preserving historic buildings, use methods that are reversible. Future generations may develop better methods for preservation.
2. Placing assumptions on top of other assumptions will skew the solution. Carefully analyze the information and look for the simplest explanation of behavior.
3. Consider several solutions and trust your intuition. Also, it takes time to simplify an answer.
4. Simple structures are easier to build and cost less. If the project is difficult to design and draw, then it will be difficult, and most likely more expensive, to build.
5. The greatest savings in resources during design and construction are the result of careful planning and the consideration of viable alternatives.
6. For historic structures, consider the least invasive repair approach. Follow the Secretary of the Interior's Guidelines for the Treatment of Historic Properties. If **strength** is insufficient, then use structural augmentation in lieu of replacement.[9]

Jonathan S. Price

Figure 1.6 Dominic Piperno and Ed Goltz (left to right; Unkefer Brothers Construction Co.) standing next to the truss

NOTES

1. "William Penn," www.pennsburymanor. org/history/william-penn/ (accessed August 6, 2016).

2. D. H. Develin, "Some Historical Spots in Lower Merion Township," Daughters of the American Revolution, 1906.

3. "Brief History of William Penn," www.ushistory.org/penn/bio.htm (accessed August 6, 2016).

4. D. M. Facenda, "Merion Friends Meetinghouse: Documentation and Site Analysis," Thesis (Master of Science in Historic Preservation, University of Pennsylvania, 2002).

5. D. T. Yeomans, "A Preliminary Study of 'English' Roofs in Colonial America," *Association for Preservation Technology Bulletin* 13, no. 4 (1981): 16. Yeomans acknowledges the assistance of Batcheler in discussing the Merion roof.

6. Acts and Ordinances of the Interregnum, 1642–1660, www.british-history.ac.uk/no-series/acts-ordinances-interregnum (accessed October 27, 2016).

7. Keast & Hood, *Merion Friends Structural Assessment Report* (Philadelphia, PA: Keast & Hood) 2005.

8. Keast & Hood, *Merion Friends Structural Assessment Report* (Philadelphia, PA: Keast & Hood) 2005.

9. K. D. Weeks and A. E. Grimmer, *The Secretary of the Interior's Standards for the Treatment of Historic Properties* (Washington, DC: National Park Service) 1995.

Timber Fundamentals

Chapter 2

Paul W. McMullin

Timber is the most versatile structural material. It has been used for centuries because of its availability, workability, and light weight. We use it in formwork when we cast concrete, for housing, retail, religious, and numerous other structures. Today, we use light and heavy **sawn lumber** and **engineered wood** products, made from smaller wood pieces. Timber structure heights are commonly in the one-to-five-story range. However, engineers and architects are developing taller wood structures, given timber's high strength-to-weight ratio, advances in engineered wood products, and sustainable characteristics. Some say this is the century of timber construction.

2.1 HISTORICAL OVERVIEW

Nomadic hunter-gatherers were the first to use timber in light, transportable tent structures; essential for their survival. A number of these types of structures are still in use today. They include yurts in Asia, teepees in North America, and goat-hair tents in the Middle East and North Africa.

As humankind became increasingly agrarian, they established permanent settlements. Metal tools provided craftsmen the means to fell large trees and work them into usable lumber. Ancient cultures used large timbers in post and beam construction that served as a skeletal framework between which non-load bearing walls and roofs were erected. Norwegian churches are a sophisticated example of this type of heavy timber construction [Figure 2.1].

Producers standardized dimensional lumber between the mid-1800s through mid-1900s. During this period, heavy timbers were replaced by balloon framing, and then by platform framing [see Figure 2.2]—the system most commonly used today. Through centuries of practice, trial and error, and the advent of modern structural engineering the art of wood construction has led to more sophisticated framing techniques, stronger engineered wood products, and more durable buildings.[1]

2.2 CODES AND STANDARDS

Many countries have their own wood design code, typically referenced by the model building code. The *National Design Specification* (***NDS***)[2] is the wood design code in the United States. Canadian engineers use *Engineering Design in Wood* by the CSA Group,[3] and European engineers use *Eurocode 5: Design of timber structures*.[4] Regardless of origin, these codes provide the minimum standard of care and a consistent technical

Paul W. McMullin

Figure 2.1 Heddal Stave Church, c. thirteenth century, Heddal, Notodden, Norway
Source: Image courtesy Teran Mitchell

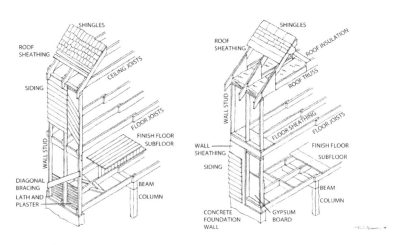

Figure 2.2 Balloon and platform framing
Source: Image courtesy Teran Mitchell

basis for designing timber structures. Table 2.1 summarizes the *NDS*—
the basis of this book—in case you desire to study a given topic further.

2.2.1 ASD Methodology

There are two timber design philosophies: allowable stress design (ASD)
and load and resistance factor design (**LRFD**). We will use **ASD** in this
book, as the *NDS* is a stress-based methodology, whereas LRFD is
capacity-based. To convert to LRFD, we use factored loads and apply a
few **adjustment factors**.

In ASD, we calculate member stress and compare it with the allowable
stress, using units of lb/in^2 (kN/m^2). Material **reference design values**
include **safety factors** to account for unknowns in material strength,
stiffness, and loading. Additionally, we apply *NDS*-specified adjustment
factors to adjust for influences such as moisture, temperature, and size—
to name a few.

Safety factors provide a reasonable margin between strength and load
variation, as schematically shown in Figure 2.3. Allowable reference
design values, presented in Appendices 2 and 3, are derived from the
ultimate stress values divided by safety factors. This shifts the strength
curve to the left, while the load curve stays in the same place.
Proportioning of members is based on how close these curves get to each
other. Where they overlap, failure could occur—the probability is often
expressed as 1/10,000; in other words, there is a 0.01 percent chance of
failure in a properly designed structure.

You might think it would then make sense to increase the margin of
safety between **demand** and capacity to eliminate the possibility of failure.
This would be good if no one had to pay for the materials, or if our
environment didn't have to support their extraction and manufacturing. As
both are driving concerns on projects, we balance demand and capacity
against risk, cost, schedule, and environmental constraints. This is the art
of engineering.

The curve shape in Figure 2.3 is a function of variability. Narrow curves
have less deviation than wider curves. Figure 2.4 shows this strength
distribution for sawn lumber, **glued laminated timber**, and **structural
composite lumber** (SCL).

2.2.2 Height and Area Limits

The *International Building Code* (*IBC*) currently limits timber building
heights and areas to those listed in Table 2.2. The maximum height is

Paul W. McMullin

Table 2.1 *NDS* Timber code summary

Chapter	Title	Contents
1	General Requirements for Structural Design	Scope and basic information related to using the code
2	Design Values for Structural Members	Introduction of reference strength and adjustment factors
3	Design Provisions and Equations	Basic bending, shear and compression equations
4	Sawn Lumber	Provisions for sawn lumber
5	Structural Glued Laminated Timber	Provisions for glued laminated timber
6	Round Timber Poles and Piles	Provisions for poles and piles
7	Prefabricated Wood I-Joists	Provisions for I-joists
8	Structural Composite Lumber	Provisions for composite lumber
9	Wood Structural Panels	Provisions for structural panels (plywood)
10	Cross-Laminated Timber	Provisions for cross-laminated timber
11	Mechanical Connections	Provisions for mechanical connections
12	Dowel-Type Fasteners	Reference strengths for bolts, lag screws, and nails
13	Split Ring and Shear Plate Connectors	Reference strengths for split ring and shear plate connectors
14	Timber Rivets	Reference strengths for timber rivets
15	Special Loading Conditions	Provisions for floor load distribution and built-up columns
16	Fire Design of Wood Members	Fire design guidance for members and connections
A	Construction and Design Practices	Specification-like requirements for design and construction
B	Load Duration	Expanded information on load duration CD factors

Table 2.1 *continued*

Chapter	Title	Contents
C	Temperature Effects	Expanded guidance on temperature considerations
D	Lateral Stability of Beams	Expanded information on beam stability CL factors
E	Local Stresses in Fastener Groups	Additional guidance on local fastener failure modes
F	Design for Creep and Critical Deflection Applications	Information regarding creep design
G	Effective Column Length	Column length factor table
H	Lateral Stability of Columns	Expanded information on column stability CP factors
I	Yield Limit Equations for Connections	Expands on connection failure modes
J	Solution of Hankinson Formula	Solutions when bearing stresses are at an angle to the grain
K	Typical Dimensions for Split Ring and Shear Plate Connectors	Dimensions for split ring and shear plates
L	Typical Dimensions for Dowel-Type Fasteners and Washers	Dimensions for bolts, lag screws, and nails
M	Manufacturing Tolerances for Rivets and Steel Side Plates for Timber Rivet Connections	Dimensions for rivets and steel plates
N	Load and Resistance Factor Design (LRFD)	Guidance on converting to LRFD

Source: NDS 2015

Paul W. McMullin

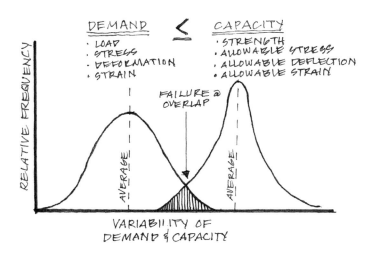

Figure 2.3 Strength and load distribution

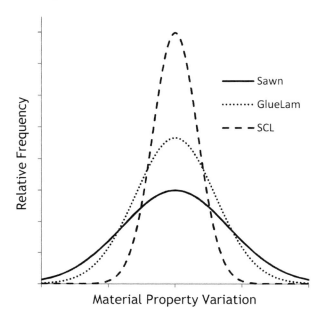

Figure 2.4 Material property variation for different wood products

Table 2.2 Representative timber-building height and area limitations

Occupancy Group	Type III		Type IV	Type V	
	A	B	HT	A	B
Maximum Height, ft (m)					
All	65 (19.8)	55 (16.8)	65 (19.8)	50 (15.2)	40 (12.2)
Permitted Stories **Permitted Area, ft² (m²)**					
A-1, Assembly	3 14,000 (1,301)	2 8,500 (790)	3 15,000 (1,394)	2 11,500 (1,068)	1 5,500 (511)
A5, Assembly	UL UL	UL UL	UL UL	UL UL	UL UL
B, Business	5 28,500 (2,648)	3 19,000 (1,765)	5 36,000 (3,345)	3 18,000 (1,672)	2 9,000 (836)
E, Education	3 23,500 (2,183)	2 14,500 (1,347)	3 25,500 (2,369)	1 18,500 (1,719)	1 9,500 (883)
F-1, Factory	3 19,000 (1,765)	2 12,000 (1,115)	4 33,500 (3,112)	2 14,000 (1,301)	1 8,500 (790)
H-1, High Hazard	1 9,500 (883)	1 7,000 (650)	1 10,500 (975)	1 7,500 (697)	NP NP (NP)
H-4, High Hazard	5 28,500 (2,648)	3 17,500 (1,626)	5 36,000 (3,345)	3 18,000 (1,672)	2 6,500 (604)
I-2, Institutional	1 12,000 (1,115)	NP NP (NP)	1 12,000 (1,115)	1 9,500 (883)	NP NP (NP)
M, Mercantile	4 18,500 (1,719)	2 12,500 (1,161)	4 20,500 (1,905)	3 14,000 (1,301)	1 9,000 (836)

Paul W. McMullin

Table 2.2 *continued*

R1, 2, 4, Residential	4 UL	4 UL	4 UL	3 UL	3 UL
S-1, Storage	3 26,000 (2,415)	2 17,500 (1,626)	4 25,500 (2,369)	3 14,000 (1,301)	1 9,000 (836)
U, Utility	3 14,000 (1,301)	2 8,500 (790)	4 18,000 (1,672)	2 9,000 (836)	1 5,500 (511)

Notes: UL = unlimited; NP = not permitted
Source: IBC 2012

65 ft (20 m), the maximum number of stories is 5, and the area is 36,000 ft² (3,345 m²). These limits are based on fire rating requirements, not strength. An effort is underway to extend timber use to taller structures,[5] focusing on composite lumber products. The Canadian building code recently allowed six stories in wood construction. Timber buildings of 10–18 stories are beginning to be constructed or planned in Australia and Europe.

2.3 MATERIALS

Wood consists of **cellulose** and **lignin**, illustrated in Figure 2.5. Long cellulose fibers give it strength along the grain, while lignin holds the fibers together, providing shear strength and load transfer between discontinuous fibers. The fibers form the strongest part of this matrix.

Timber materials consist of sawn or engineered (manufactured) lumber. Engineered lumber includes glued laminated timber, SCL, I-joists, and structural **panels**.

2.3.1 Solid Sawn Lumber

Solid sawn lumber is milled from harvested trees, whereas engineered wood products are made from smaller wood pieces glued together. Sawn lumber comes in three size types: full sawn, rough sawn, and **dressed** (see Figure 2.6). Dressed lumber is the commonest type and what we buy at the lumber yard (Figure 2.7). Nominal sizes include 2 × 4, 2 × 8, and 4 × 4. For rough-cut lumber, the nominal and actual sizes were historically the same, but then, after 1870, when surfacing became the norm, lumber-producing organizations agreed on finished dimensions for standardization.[6] Today, lumber is ½–¾ in (12–19 mm) smaller than the nominal (specified) size.

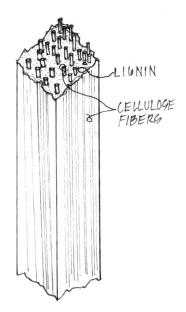

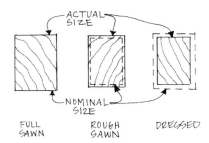

Figure 2.5 Conceptual sketch of lignin and cellulose

Figure 2.6 Solid sawn lumber size types, after Fridley[7]

These sizes are listed in Table A1.1 in Appendix 1 for common dressed shapes.

Representative sawn lumber reference design values (strength and stiffness) are provided in Appendix 2 in the following tables (refer to the *NDS* for additional species and grades):

- visually graded **dimension lumber**—Table A2.1;
- visually graded timbers (5×5 in and larger)—Table A2.2;
- mechanically graded Douglas Fir–Larch (North)—Table A2.3;
- visually graded Southern Pine—Table A2.4.

These tables contain the reference design strength and stiffness. To find **adjusted design values**, we multiply these by the adjustment factors specific to sawn lumber, listed in Table 2.3, and, more specifically, for sawn lumber, in Table 2.4—discussed further in Section 2.4. Note that all the available adjustment factors do not apply all the time.

Paul W. McMullin

Figure 2.7 Dressed lumber in porch canopy

2.3.2 Engineered Wood

Engineered wood products are substantially stronger, have greater stiffness, and increase the sustainability of timber. They are making longer spans and taller structures possible. Engineered products include glued laminated timber, SCL, I-joists, and structural panels, shown in Figure 2.8. The manufacturers place stronger and greater amounts of wood in areas of highest stress, reducing overall material use. For example, I-joists are designed as bending members. The flanges are larger and made of higher-strength material. The web is thinner, but made of structural panels, which are efficient in shear. The net result is less material use.

Table 2.3 Adjustment factor location guide

Factor				Material Type					Note
Use	Symbol	Section	Table or Equation	Sawn Lumber	Glued Laminated Timber	Structural Composite Lumber (SCL)	I-Joists	Connections	
Load duration	C_D	App 4	Table A4.1	X	X	X	X	X	ASD only
Wet service	C_M	App 4	Tables A4.2 & A4.3	X	X	X	X	X	Varies for each wood table
Temperature	C_t	App 4	Table A4.4	X	X	X	X	X	
Beam stability	C_L	Ch 4	Equation (4.1)	X	X	X	X		
Size	C_F	App 4	Table A4.5	X					Varies for each strength table
Volume	C_V	App 4	Tables A4.6 & A4.7		X	X			Varies for glued laminated and SCL
Flat use	C_{fu}	App 4	Table A4.8	X	X				Varies for each strength table

				<div style="writing-mode:vertical-rl">Sawn</div>	<div>GLT</div>	<div>SCL</div>	<div>I-joist</div>	<div>Connections</div>	
Curvature	C_c	App 4	Table A4.9		X				
Stress interaction	C_I	Ch 4	Ch 2 eqn.		X				
Shear reduction	C_{vr}	Ch 5	Ch 2 eqn.		X				
Incising	C_i	App 4	Table A4.10	X					
Repetitive member	C_r	Ch 2	Section 2.4.11	X		X	X		
Column stability	C_P	Ch 6	Ch 5 eqn.	X	X	X			
Buckling stiffness	C_T	Ch 2	Section 2.4.13	X					
Bearing area	C_b	App 4	Table A4.11	X	X	X			
Group action	C_g	App 4	Table A4.12					X	
Geometry	$C\Delta$	App 4	Tables A4.13 & A4.14					X	
End grain	C_{eg}	App 4	Table A4.15					X	
Diaphragm	C_{di}	Ch 9	Section 9.2.1.b					X	
Toe-nail	C_{tn}	App 4	Table A4.16					X	
Format conversion	K_F	App 4	Table A4.17	X	X	X	X	X	LRFD only
Resistance	ϕ	App 4	Table A4.18	X	X	X	X	X	LRFD only
Time effect	λ	App 4	Table A4.19	X	X	X	X	X	LRFD only

Table 2.4 Sawn lumber adjustment factors

Action	Equation		ASD only — Load Duration	ASD & LRFD — Wet Service	Temperature	Beam Stability	Size	Flat Use	Incising	Repetitive Member	Column Stability	Buckling Stiffness	Bearing Area	LRFD only — K_F Format Conversion	ϕ Resistance	Time Effect
Bending	$F'_b = F_b$	×	C_D	C_M	C_t	C_L	C_F	C_{fu}	C_i	C_r				2.54	0.85	λ
Tension	$F'_t = F_t$	×	C_D	C_M	C_t		C_F		C_i					2.70	0.80	λ
Shear	$F'_v = F_v$	×	C_D	C_M	C_t				C_i					2.88	0.75	λ
Compression ∥	$F'_c = F_c$	×	C_D	C_M	C_t		C_F		C_i		C_P			2.40	0.90	λ
Compression ⊥	$F'_{c\perp} = F_{c\perp}$	×		C_M	C_t				C_i				C_b	1.67	0.90	
Elastic Modulus	$E' = E$	×		C_M	C_t				C_i							
Min Elastic Modulus	$E'_{min} = E_{min}$	×		C_M	C_t				C_i			C_T		1.76	0.85	

Source: NDS 2015

Figure 2.8 Engineered wood products. From top left, clockwise: glued laminated timber, PSL, LVL, I-joist, and structural panel

Advantages of engineered wood include:

- It is made using smaller trees: This reduces the growing time for replanted forests and encourages sustainable forestry practices.
- There are fewer **discontinuities**: Because the **component** pieces are small and glued together, the flaws are inherently smaller.
- There is material where we need it. The 'engineered' in wood products refers to the shape and material strength distribution being designed to best resist stresses. This results in stronger, more efficient members.

Disadvantages include:

- Limited cross-application: Some members are designed for a specific application. For example, I-joists are not able to function as **columns**.
- Water susceptibility: The glues in many engineered wood products are not able to withstand long periods of water exposure.
- Treatment averse: The glues in engineered wood products are often not compatible with preservative or fire treatments.

2.3.2.a Glued Laminated

Glued laminated timbers are made of nominal 2× material, glued together flatwise (Figure 2.9), thereby making larger members. Manufacturers economize by placing higher-quality material in the top and bottom laminations (lams or layers), where bending stress is greatest, and lower-quality, less-expensive woods for the middle lams, where shear is dominating. Glued laminated members are available in a variety of curved shapes, making them popular for gymnasiums, natatoriums, and places of worship.

Table A2.5 in Appendix 2 contains common glued laminated timber reference design stress and stiffness values. To find adjusted design values, multiply these by the adjustment factors specific to glued laminated timber, listed in Table 2.5.

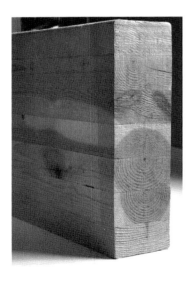

Figure 2.9 Glued laminated beam cross section. Sample courtesy Boise Cascade

Paul W. McMullin

Table 2.5 Glued laminated timber adjustment factors

Action	Equation		ASD only	ASD & LRFD										LRFD only		
			Load Duration	Wet Service	Temperature	Beam Stability	Volume	Flat Use	Curvature	Stress Interaction	Shear Reduction	Column Stability	Bearing Area	Format Conversion K_F	Resistance ϕ	Time Effect
Bending	$F'_b = F_b$	x	C_D	C_M	C_t	C_L	C_V	C_{fu}	C_c	C_I				2.54	0.85	λ
Tension	$F'_t = F_t$	x	C_D	C_M	C_t									2.70	0.80	λ
Shear	$F'_v = F_v$	x	C_D	C_M	C_t						C_{vr}			2.88	0.75	λ
Radial Tension	$F'_{rt} = F_{rt}$	x	C_D	C_M	C_t									2.88	0.75	λ
Compression $\parallel$	$F'_c = F_c$	x	C_D	C_M	C_t							C_P		2.40	0.90	λ
Compression $\perp$	$F'_{c\perp} = F_{c\perp}$	x		C_M	C_t								C_b	1.67	0.90	
Elastic Modulus	$E' = E$	x		C_M	C_t											
Min Elastic Modulus	$E'_{min} = E_{min}$	x		C_M	C_t									1.76	0.85	

Source: NDS 2015

Note the F_{b+} and F_{b-} values in the second and third columns are sometimes the same and other times different. Identical values indicate a balanced **layup**, whereas different values indicate an unbalanced layup. The difference stems from the intended use. Unbalanced layups are for simple span beams, where there will only be tension in the bottom lams. Manufacturers place higher-quality wood in the bottom lams, because tension allowable stress is lower than compression allowable stress for the same species and grade. A good way to tell if the beam is installed correctly is to look at the bottom. If it says 'up', the contractor has installed it upside down. In this case, we use the F_{b-} values to evaluate the beam. Balanced layups are intended for multi-span beams where there will be tension in the top lams over the supports. It doesn't matter which way they are installed.

2.3.2.b Structural Composite Lumber

SCL is made of layers or strands of wood, glued together. It encompasses four product types: laminated veneer lumber (**LVL**), parallel strand lumber (**PSL**), laminated strand lumber (**LSL**), and cross laminated timber (CLT). SCL is often used to replace solid sawn joists and light beams and, in recent years, has become available in column form.

We use LVL for beams, joists, **headers**, and for **I-joist** flanges. Resembling plywood, it is thicker, and the lams are all parallel, as shown

Figure 2.10 Laminated veneer lumber (LVL) cross section

Paul W. McMullin

in Figure 2.10. It comes in 1¾ in (44.5 mm) thicknesses and **depths** ranging from 5½ to 18 in (140–457 mm), as summarized in Table A1.3 in Appendix 1.

PSL consists of narrow wood strips, cut into lengths and having a strand length-to-thickness ratio of about 300, which are laid parallel and glued together, as shown in Figure 2.11. PSLs are used where high axial strengths are required. They are available in sizes ranging from 3½ to 7 in (89–180 mm) wide and from 3½ to 18 in (89–460 mm) deep (see Table A1.3 in Appendix 1). Check with your local supply house for size availability.

LSL is made from flaked wood strands, having a length-to-thickness ratio around 150, which are glued together as shown in Figure 2.12. We use it for studs, joists, and lower-load headers and columns. They are available in sizes ranging from 3½ to 7 in (89–180 mm) wide and from 3½ to 18 in (89–460 mm) deep.

Figure 2.11 Parallel strand lumber (PSL) in beam and column application
Source: Photo courtesy of Weyerhaeuser©

Figure 2.12 Laminated strand lumber (LSL) in header application
Source: Photo courtesy of Weyerhaeuser©

Calculations with SCL require section properties, provided in Table A1.3, Appendix 1, and reference design values, summarized in Table A2.6. These are based on calculation and manufacturers' data. Table 2.6 lists adjustment factors, which don't vary with SCL type.

Cross-laminated timber is made from alternating layers of solid sawn lumber, illustrated in Figure 2.13. They are either glued together or connected with dovetails.

2.3.2.c I-Joists

I-joists are specifically designed to replace solid sawn 2× members used as repetitive floor joists or roof rafters. They are constructed of oriented strand board (OSB) or plywood webs with LVL or solid sawn flanges

Paul W. McMullin

Table 2.6 Structural composite lumber adjustment factors

Action	Equation		ASD only	ASD & LRFD							LRFD only		
			Load Duration	Wet Service	Temperature	Beam Stability	Volume	Repetitive Member	Column Stability	Bearing Area	K_F Format Conversion	ϕ Resistance	Time Effect
Bending	$F'_b = F_b$	×	C_D	C_M	C_t	C_L	C_V	C_r			2.54	0.85	λ
Tension	$F'_t = F_t$	×	C_D	C_M	C_t						2.70	0.80	λ
Shear	$F'_v = F_v$	×	C_D	C_M	C_t						2.88	0.75	λ
Compression ∥	$F'_c = F_c$	×	C_D	C_M	C_t				C_P		2.40	0.90	λ
Compression ⊥	$F'_{c\perp} = F_{c\perp}$	×		C_M	C_t					C_b	1.67	0.90	
Elastic Modulus	$E' = E$	×		C_M	C_t								
Min Elastic Modulus	$E'_{min} = E_{min}$	×		C_M	C_t						1.76	0.85	

Source: NDS 2015

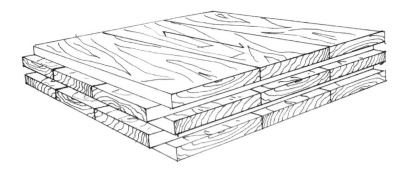

Figure 2.13 Cross-laminated timber
Source: Courtesy COCIS, Edinburgh Napier University, & ITAC, University of Utah

(see Figure 2.14). Joist depths range from 9½ to 20 in (240–510 mm). Bending stresses are highest at the top and bottom, where the strongest material in the joist is placed. Section and strength properties are combined into bending, shear, and bearing strengths, as summarized in Table A2.7, in Appendix 2. This is unlike the other materials where we use stress and section properties in our calculations. Table 2.7 lists the adjustment factors that are specific to I-joists.

Figure 2.14 I-joist cross section with LVL flanges and OSB web

Paul W. McMullin

Table 2.7 I-joist adjustment factors

Action	Equation		ASD only Load Duration	ASD & LRFD Wet Service	Temperature	Beam Stability	Repetitive Member	LRFD only K_F Format Conversion	ϕ Resistance	Time Effect
Bending	$M'_r = M_r$	x	C_D	C_M	C_t	C_L	C_r	K_F	0.85	λ
Shear	$V'_r = V_r$	x	C_D	C_M	C_t			K_F	0.75	λ
Reaction	$R'_r = R_r$	x	C_D	C_M	C_t			K_F	0.75	λ
Stiffness	$E_I' = E_I$	x		C_M	C_t					
Min Stiffness	$(EI)'_{min} = (EI)_{min}$	x		C_M	C_t			K_F	0.85	
Shear Stiff	$K' = K$	x		C_M	C_t					

Source: NDS 2015

2.3.2.d Structural Panels

We use structural panels as roof, floor, and wall sheathing. Panels must meet one of two standards: APA PS 1 (plywood) or PS 2 (plywood and OSB). Panels carry **gravity loads** through bending, and shear from lateral loads. They are not effective at carrying tension or compression. Plywood is made from thin layers (veneers) of wood, each at 90° to the previous and glued together (see Figure 2.15). **Oriented strand board** is made from strand-like chips that are oriented more along the length of the panel and glued together, as shown in Figure 2.16.

2.3.3 Connectors

Connections between members are fundamental to timber's successful use. Without them, a timber structure would amount to little more than a pile of sticks. Historically, builders used mortise and tenon, dovetail, dowel, split ring, and shear plate to connect wood members. Connectors today include **nails**, bolts, **lag screws**, **truss plates**, timber rivets, and

Figure 2.15 Plywood sheet showing alternating ply directions

Figure 2.16 Oriented strand board (OSB) showing primary strand direction

Paul W. McMullin

engineered **metal plate connectors**, although the older joints still perform well and provide an interesting flair to an exposed wood project.

This text covers common **dowel-type** fasteners and engineered metal plate connectors. Dowel-type fasteners consist of nails, bolts, and lag screws, shown in Figure 2.17. They transfer force either in bearing (i.e., shear) or tension (withdrawal). Engineered connectors are extremely varied and provide versatility for timber framing; see Figure 2.18. Simpson Strong-Tie™ and USP™ produce the greatest number and variety of proprietary connectors.

We find reference design strengths for dowel-type connectors in Appendix 3, applicable adjustment factors in Table 2.8, and expanded information in Chapter 9. Proprietary **connection** reference design strengths are in the manufacturer literature or, preferably, International Code Council Evaluation Service Reports (ICC ESR).

2.4 ADJUSTMENT FACTORS, C_{xx}

Adjustment factors account for various influences on strength and stiffness. This allows us to use one set of reference design strengths values (see Appendix 3) and then modify them for external variables.

Figure 2.17 Dowel-type fasteners, showing, top to bottom, nails, lag screws, and bolts

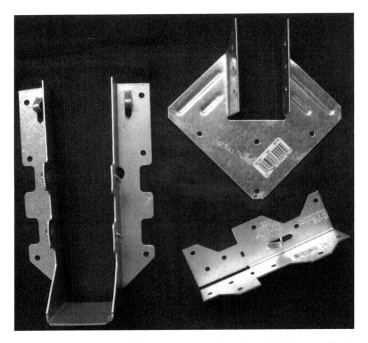

Figure 2.18 Engineered metal plate connectors showing a joist hanger, hurricane tie, and A35 clip

We multiply the *NDS* reference design values by all applicable factors. Some are greater than 1.0; many are 1.0 or less, depending on the service conditions.

Determination of the adjustment factors is the most time-consuming effort in timber design. Take your time to become familiar with their location in this book. Table 2.3 lists all the adjustment factors, their applicability, and where they are discussed in the text and tabulated in Appendix 4. A brief discussion of each factor follows in the next sections.

2.4.1 Load Duration, C_D—Table A4.1

Load duration factors account for how long a load will be carried by the member. They range from 0.9 for **dead** (permanent) to 2.0 for impact loads.

Table 2.8 Connector adjustment factors

Action	Equation	ASD only	ASD & LRFD							LRFD only		
		Load Duration	Wet Service	Temperature	Group Action	Geometry	End Grain	Diaphragm	Toe-Nail	Format Conversion	Resistance	Time Effect
Lateral Loads Dowel-type Fasteners	$Z' = Z \times$	C_D	C_M	C_t	C_g	C_Δ	C_{eg}	C_{di}	C_{tn}	K_F 3.32	ϕ 0.65	λ
Withdrawal Loads Nails, spikes, lag screws, wood screws & drift pins	$W' = W \times$	C_D	C_M	C_t			C_{eg}		C_{tn}	3.32	0.65	λ

Note: For engineered metal plate connectors, see ASTMD5457

Source: NDS 2015

2.4.2 Wet Service, C_M—Tables A4.2 and A4.3

The **wet service factor** captures the structural effect of high moisture conditions. It is 1.0 for interior use and moisture content (less than or equal to 19 percent). The factors vary for different lumber species and sizes for moisture content greater than 19 percent. Typical interior conditions are considered dry.

2.4.3 Temperature, C_t —Table A4.4

Temperature factors reflect the influence of sustained, elevated temperatures above 100°F (38°C). They are typically 1.0, as most structures don't experience high, sustained temperatures, but C_t can be as low as 0.5 for service in wet environments at temperatures between 125°F and 150°F. The factors are different for wood and connectors and vary with moisture content.

2.4.4 Beam Stability, C_L

Beam stability factors account for the tendency of tall, narrow beams to roll over in the middle. They are determined by equation and further discussed in Chapter 4.

2.4.5 Size, C_F, and Volume, C_V—Tables A4.5–A4.7

As the size of a member increases, so does the potential for discontinuities, such as knots. **Size** and **volume factors** adjust for this; they range from 0.4 to 1.5 and vary between different types and species of lumber.

2.4.6 Flat Use, C_{fu}—Table A4.8

Flat use factors account for the strength and stiffness variation when the member is loaded along its wide face (laid flatwise). They vary from 0.74 to 1.2. These factors commonly apply to timber decking.

2.4.7 Curvature, C_c—Table A4.9

The **curvature factor** accounts for additional stresses that occur in the manufacturing process of curved, glued laminated beams. It is calculated by equation in the *NDS*.[8]

2.4.8 Stress Interaction, C_I

Stress interaction factors account for taper in glued laminated bending members. They are outside the scope of this book, but are calculated by equations provided in the code.[9]

Paul W. McMullin

2.4.9 Shear Reduction, C_{vr}

Shear reduction factors apply to glued laminated timber. The factor is 0.72 for non-prismatic members, members with impact load, members with notches, and at connections.

2.4.10 Incising Factor, C_i—Table A4.10

The **incising factor** accounts for little cuts made into wood to allow preservatives and fire retardants to be injected (see Figure 2.19). Incising cuts are ⅜ in (10 mm) deep and long, and run parallel to the grain.

2.4.11 Repetitive Member, C_r

The **repetitive member factor** accounts for the potential load sharing between closely spaced members (24 in (600 mm) or less on center). They must be connected with a load sharing element, such as floor sheathing. The repetitive member factor is 1.15 for sawn lumber and 1.04 for SCL. It is 1.0 for all other materials, including I-joists.

2.4.12 Column Stability, C_P

Column stability factors account for the **buckling** potential of **slender** columns. They are determined by a nonlinear equation based on $(L/d)^2$, where L is column height, and d is least dimension of the cross section—further discussed in Chapter 5.

2.4.13 Buckling Stiffness, C_T

The **buckling stiffness factor** applies only to 2 × 4 sawn lumber truss **chords** in combined bending and compression. It is determined by equation from the *NDS*.[10]

Figure 2.19 Incising cuts in pressure treated wood

2.1.14 Bearing Area, C_b—Table A4.11

The **bearing area factor** increases compressive strength perpendicular to the grain when the load is applied 3 in (76 mm) or more away from the end, as shown in Figure 2.20. It commonly applies to studs or columns bearing on a horizontal wood member. It is 1.0 for bearing lengths of 6 in (150 mm) or more.

2.4.15 Group Action, C_g—Table A4.12

Group action factors account for the nonuniform load distribution in bolt or lag screw connections with multiple fasteners (see Figure 2.21). C_g decreases as the number of bolts or screws increase.

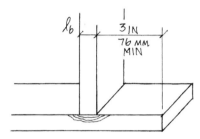

Figure 2.20 Wall stud bearing on sill plate showing applicability of the bearing adjustment factor

Figure 2.21 Bolted connection with multiple fasteners requiring application of the group adjustment factor

Paul W. McMullin

2.4.16 Connection Geometry, C_Δ—Tables A4.13 and A4.14

Geometry factors reduce the connector strength when it is located too close to an edge or end of a member, or to another connector. It is 1.0 if the minimum end, edge, and **spacing** requirements are met—see Chapter 9—but can be as low as 0.5.

2.4.17 End Grain, C_{eg} —Table A4.15

The **end grain factor** reduces the strength of dowel-type fasteners installed in the end of a member, parallel to the grain (see Figure 2.22). It ranges between 0 and 0.67, depending on fastener and load direction. Nails and wood screws in the end grain loaded in withdrawal (tension) have no strength, hence a C_{eg} factor of 0.

2.4.18 Diaphragm, C_{di}

The **diaphragm factor**—C_{di} = 1.1—increases lateral nail design values. A diaphragm is the sheathing on a floor or roof, designed to resist lateral loads (**wind** or **seismic**).

2.4.19 Toe-Nail, C_{tn}—Table A4.16

Toe-Nail factors reduce connection strength when installed diagonally through the connected member, as shown in Figure 2.23.

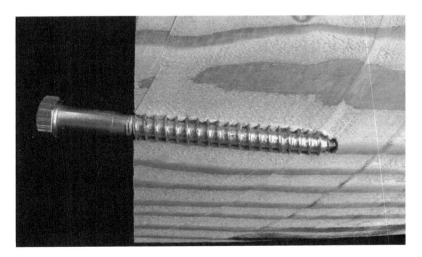

Figure 2.22 Lag screw installed in the end grain of a member

Figure 2.23 Cut-away view of a toe-nailed connection

2.4.20 Format Conversion, K_F—Table A4.17

Format conversion factors take the reference design values located in Appendices 2 and 3 from ASD to LRFD levels. They vary from 1.76 to 3.32, depending on the property type. We do not use them with ASD design.

2.4.21 Resistance, ϕ—Table A4.18

Resistance factors reduce material values used in LRFD design. They range from 0.65 to 0.90. In LRFD design, they shift material values to the left of the curve in Figure 2.3, whereas the **load factors** shift the demand curve to the right.

2.4.22 Time Effect, λ—Table A4.19

The **time effect factor** is essentially the reciprocal of the load duration factor. It ranges from 0.6 to 1.0, depending on the load type, and is only used in LRFD design.

Paul W. McMullin

We multiply the applicable adjustment factors by the reference design values given in Appendices 2 and 3. For example, we find allowable axial compressive stress F'_c of a sawn lumber column by multiplying the reference design stress F_c by the applicable adjustment factors C_D, C_M, and so on, shown as follows:

$$F'_c = F_c C_D C_M \, C_t C_i C_P \tag{2.1}$$

Each material type uses a given set of adjustment factors for each stress action (e.g., tension, bending, and compression). These are summarized in the tables listed below, found earlier in this chapter. They will be very helpful in keeping track of your calculations.

- Sawn lumber—Table 2.4;
- Glued laminated timber—Table 2.5;
- Structural composite lumber (SCL)— Table 2.6;
- I-joists—Table 2.7;
- Mechanical connectors—Table 2.8.

2.5 MATERIAL BEHAVIOR

2.5.1 Anisotropic

Timber is **anisotropic**, meaning its material properties are different in each direction. Properties in the direction of the wood fibers (grain) are highest, and they are lowest cross-wise (perpendicular) to the grain (see Figure 2.24). This anisotropy is handled by attention being paid to load direction, and by the appropriate reference design stresses being used, namely:

- F_c for compression parallel to the grain;
- $F_{c\perp}$ for compression perpendicular to the grain;
- F_t for tension parallel to the grain (we never want tension loads perpendicular to the grain: see discussion in Chapter 3).

Compression reference design stresses parallel to the grain, F_c, are two to three times greater than the strength perpendicular to the grain, $F_{c\perp}$.

2.5.2 Stress–Strain Curve

Stress–strain relationships in timber vary depending on load type and direction. For bending, stress and strain are linear up to the proportional limit, where the strain increases faster than stress until failure occurs (see Figure 2.25). For tension along the wood fibers, strain and stress increase linearly until about 0.5–1 percent strain, then abrupt failure occurs before

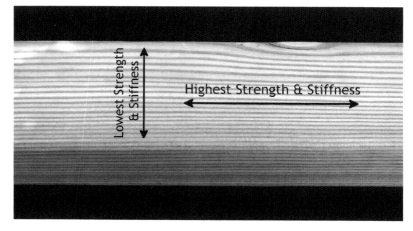

Figure 2.24 Anisotropy in timber

2 percent, with little nonlinear behavior, as shown in Figure 2.26. Under compression parallel to the grain, stress and strain increase linearly until around 0.5–1 percent strain, where it reaches the proportional limit (similar to **yielding** in a steel specimen). Compression strength slowly degrades with increasing strain, as shown in Figure 2.27. Compression perpendicular to the grain starts linearly, and then arrives asymptotically at a maximum strength, without degradation (Figure 2.28), because the wood becomes increasingly dense. Timber design stays in the linear portion of these stress–strain curves, but it can be helpful to know where reserve **capacity** may lie.

2.5.3 Discontinuities

All building materials have imperfections. Timber commonly has knots, **splits**, **checks**, and **shakes**, as shown in Figure 2.29. These are not necessarily flaws, simply discontinuities. Reference design values take this reality into account. Heavy timber beams are good examples of members containing large discontinuities that have little impact on overall strength (Figure 2.30).

2.6 SECTION PROPERTIES

The commonest section properties used in designing timber structures are area, A, **section modulus**, S, **moment of inertia**, I, and **radius of**

Paul W. McMullin

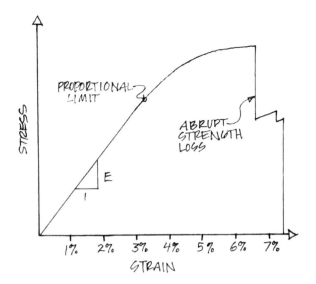

Figure 2.25 Bending stress–strain curve

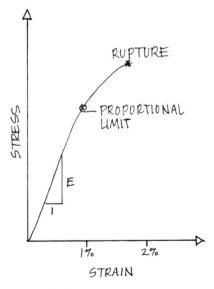

Figure 2.26 Tension stress–strain curve

Timber Fundamentals

47

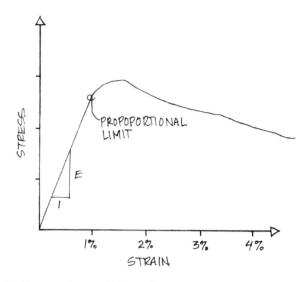

Figure 2.27 Compression parallel to grain stress–strain curve

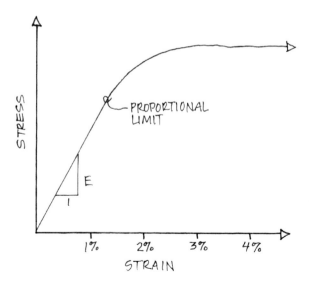

Figure 2.28 Compression perpendicular to grain stress–strain curve

Paul W. McMullin

Figure 2.29 Common timber discontinuities: (a) knot, (b) split, (c) check, and (d) shake

Figure 2.30 Beam with large shakes

gyration, r. Area and section modulus relate to strength, moment of inertia relates to stiffness, and radius of gyration relates to **stability**. Figure 2.31 shows section property equations for round and rectangular shapes. Appendix 1 provides the following rectangular section properties:

- sawn lumber—Table A1.1;
- glued laminated timber (western species)—Table A1.2;
- structural composite lumber (SCL)—Table A1.3.

2.7 STRUCTURAL CONFIGURATION

A great advantage of timber construction is its flexibility and ease of construction. It can be used in a wide range of structures and worked with simple, accessible tools. The U.S. building code limits timber construction

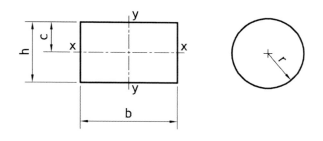

Round	Rectangle	
Area		
$A = \pi r^2$	$A = bh$	
Moment of Inertia		
$I = \dfrac{\pi r^4}{4}$	$I_x = \dfrac{1}{12}bh^3$	$I_y = \dfrac{1}{12}hb^3$
Radius of Gyration		
$r_z = \dfrac{r}{2}$	$r_x = \dfrac{h}{\sqrt{12}}$	$r_y = \dfrac{b}{\sqrt{12}}$
Section Modulus		
$S = \dfrac{\pi r^3}{4}$	$S_x = \dfrac{1}{6}bh^2$	$S_y = \dfrac{1}{6}hb^2$
$Q = \dfrac{2r^3}{3}$ at center	$Q_x = \dfrac{1}{8}bh^2$	$Q_y = \dfrac{1}{8}hb^2$
Polar Moment of Inertia		
$J = \dfrac{\pi r^4}{2}$		
$r_z = \sqrt{\dfrac{I}{A}}$ for any shape		

Figure 2.31 Section property equations

Paul W. McMullin

to five stories; the limit is six in Canada. Timber configuration is broken into light bearing wall and heavy post and beam systems.

2.7.1 Bearing Wall

Bearing wall structures are the commonest today. They are simple and efficient, as seen in Figures 2.2 and 2.32. Bearing wall systems are most effective when the walls stack—line up top to bottom—with window openings comprising less than 50 percent of the wall area. Exterior walls are often framed using 2 × 6 (50 × 150), to provide greater space for insulation. Extra attention must be paid in the design of double-height walls without an intermediate floor to brace the studs (two-story spans with large windows are particularly challenging).

2.7.2 Post and Beam

Post and beam construction is more traditional and common in historic buildings. The floor joists sit on top of beams, which sit on top of the columns, as shown in Figure 2.33. Today, heavy timber (5 in or 125 mm sections) structures are used to construct religious and public buildings and follow the historic lead of the past. They employ more modern steel connections, sometimes concealed to enhance the architectural effect.

Figure 2.32 Bearing wall residential construction

Timber Fundamentals

Figure 2.33 Post and beam in historic construction

This helps reduce the floor depth by allowing joists to frame flush to the top of beams.

2.8 CONSTRUCTION

Timber construction varies from small residential projects to large structures with prefabricated roof and wall panels. As the loads increase, we introduce engineered lumber, and then steel beams and columns where timber's strength is exceeded. Materials for residential construction are typically moved by hand or forklifts. As the project becomes larger, contractors use light cranes. Hand tools such as hammers, drills, nail guns, and circular saws are the commonest tools on the project site.

2.9 QUALITY CONTROL

The *IBC*[11] only requires special inspection on high-load diaphragm and metal plate trusses spanning 60 ft (18.3 m) or more. However, it is important for the engineer to make structural observation site visits to confirm the quality of the workmanship. In particular, attention must be paid to the following:

- Construction is following the plans, specifications, and design intent.
- The specified materials are installed.
- Correct fasteners are used.
- Nails are not overdriven, are spaced properly, and do not split the wood.
- **Blocking** is installed in the right places and is properly nailed.
- Columns are continuous through floors.
- Treated wood is installed against soil, concrete, and masonry.
- Glued laminated beams don't say 'this side up' on the bottom.
- There is coordination with electrical and mechanical trades.

2.10 WHERE WE GO FROM HERE

From here, we will dive into tension, bending, shear, and compression member design. We then get into trusses and lateral design, and end with connections.

NOTES

1. Teran Mitchell. "Structural Materials." In *Introduction to Structures*, edited by Paul W. McMullin and Jonathan S. Price (New York: Routledge, 2016).
2. AWC. *ASD/LFRD Manual, National Design Specifications for Wood Construction* (Leesburg, VA: American Wood Council, 2012).
3. CSA Group. *Engineering Design in Wood* (Toronto: CSA Group).
4. CEN. *Eurocode 5: Design of timber structures* (Brussels: CEN European Committee for Standardization).
5. "Tall Wood/Mass Timber," reTHINK WOOD, www.rethinkwood.com/masstimber (accessed May 26, 2015).
6. U.S. Department of Agriculture, Forest Service. *History of Yard Lumber Size Standards*, by L. W. Smith and L. W. Wood (Washington, DC: U.S. Government Printing Office, 1964).
7. Donald E. Breyer, Kenneth J. Fridley, and Kelly E. Cobeen. *Design of Wood Structures: ASD* (New York: McGraw-Hill, 2014).
8. ANSI/AWC. *National Design Specification (NDS) for Wood Construction* (Leesburg, VA: AWC, 2015), 39.
9. ANSI/AWC. *National Design Specification (NDS) for Wood Construction*, 39.
10. ANSI/AWC. *National Design Specification (NDS) for Wood Construction*, 32.
11. IBC. *International Building Code* (Washington, DC: International Code Council, 2012).

Timber Tension

Chapter 3

Jonathan S. Price

Humans have found shelter in timber-framed structures for millennia. These structures were built to resist gravity loads, and so rarely were any parts of the structure in tension. In contrast, modern trusses and lateral bracing systems rely on timber's substantial tensile strength.

Surprisingly, timber's strength-to-weight ratio rivals that of structural steel—although mobilizing this strength at connections remains a challenge. In 1937, Trautwine[1] reported that timber's ultimate tensile strength ranged from 6,000 lb/in^2 (41 MN/m^2) for Cypress to 23,000 lb/in^2 (159 MN/m^2) for Lancewood, although he cautioned these were breaking stress values under ideal conditions.

Timber tension members are found in trusses, hangers, and cross-bracing, as shown in Figure 3.1.

3.1 STABILITY

Stability is not a concern for tension members. The tension stresses keep the member straight and stable.

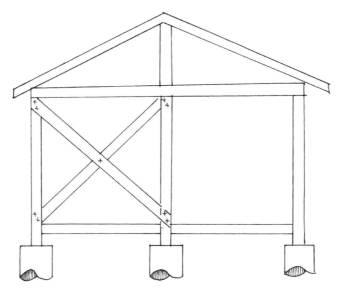

CROSS-BRACING IN TIMBER FRAME CONSTRUCTION

Figure 3.1 Timber frame construction bracing system

3.2 CAPACITY

3.2.1 Reference Design Values

Because wood fibers are arranged parallel or nearly parallel with the material's long direction, timber is **orthotropic** (Figure 3.2)—its properties change with grain direction. Timber's strength is greater parallel to the grain than perpendicular to it. Also, drying shrinkage is much greater across the grain than parallel. Drying shrinkage cause cracks that reduce bending and shear strength, but these are less of a problem for tension members.

The allowed stresses are specified by the code reference *National Design Specification* (*NDS*).[2] The stresses permitted for tension are lower than those allowed for compression, because the occasional defects, such as knots, cannot carry tension. No code safety factors account for major physical damage caused by careless handling, insect infestation, rot, or decay. This aside, carefully proportioned members are normally safe, but are vulnerable at connections.

The *NDS* publishes the reference design values for members in tension parallel to grain, F_t, bending, F_b, shear parallel to grain, F_v, compression parallel and perpendicular to grain, F_c and $F_{c\perp}$, respectively, plus **elastic modulus**, E, for commonly used species of wood. It warns us against putting timber in tension across the grain, like the condition shown in Figure 3.3, but the code adds a brief note stating, "if it is unavoidable, mechanical reinforcement shall be considered." Connections that are perpendicular to the grain unavoidably produce tension between the fibers, and so the *NDS* specifies minimum **edge distances** to avoid breakout of fasteners—this is further discussed in Chapter 9.

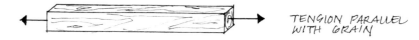

TENSION PARALLEL WITH GRAIN

Figure 3.2 Timber element in tension (parallel to grain)

TENSION PERPENDICULAR TO GRAIN

Figure 3.3 Tension across grain (not permitted)

Jonathan S. Price

The *NDS* reference design values incorporate a factor of safety (FS or SF) to account for natural defects although these factors do not compensate for design errors.

3.2.2 Adjusted Design Values

We adjust the reference design values for duration of load, wet environments, high temperature, size, and incising, as discussed in Chapter 2. Following Tables 2.4–2.7, we can quickly see which factors apply to tension for different wood products. The equations for adjusted design stress are as follows:

$$F'_t = F_t C_D C_M C_t C_F C_i \text{ for sawn lumber} \tag{3.1}$$

$$F'_t = F_t C_D C_M C_t \text{ for glued laminated timber} \tag{3.2}$$

$$F'_t = F_t C_D C_M C_t \text{ for structural composite lumber} \tag{3.3}$$

where:

F_t = reference tension design stress from Appendix 2

C_x = adjustment factors; see Table 2.3

3.2.3 Member Proportioning

To size tension members, there is one simple equation for the required net **cross-sectional area**, A_{net}—which is the gross area minus any holes made for connections:

$$A_{net} = \frac{T}{F'_t} \tag{3.4}$$

where:

T = tension force in k/in^2 (kN/m^2)

F'_t = adjusted design stress in lb (kN)

3.3 DEMAND VS. CAPACITY

The *NDS* reference design value for basic allowable tension stress is designated F_t. After it has been multiplied by the adjustment factors (C_D, C_M, C_t, C_F, and C_i), the stress to be used for structural design f_t is determined. This is also known as the adjusted design stress. We compare this stress with the actual stress in the member, calculated using:

$$f_t = \frac{T}{A_{net}} \tag{3.5}$$

INITIAL TENSION MEMBER SIZING

Rules of thumb, also known as **empirical** design, extend back some two millennia. Little has changed our appreciation for heavy post and beam construction proportioned on rules of thumb.

For tension members, a good rule of thumb is to double the required area so that lost material at connections is accommodated.

Table 3.1 Tension strength for varying species and wood types

Imperial Units										
		Tension Strength, T (k)								
	b	d	Bald Cypress	Spruce-Pine-Fir	Red-wood	Douglas Fir–Larch (N)	Glued Lamin-ated	Southern Pine	LVL	LVL
			F_t (lb/in²)							
	in	*in*	*425*	*550*	*800*	*1,000*	*1,100*	*1,350*	*1,555*	*2,485*
2×4	1.5	3.5	1.49	1.93	2.80	3.50	3.85	4.73	5.44	8.70
2×6	1.5	5.5	2.34	3.03	4.40	5.50	6.05	7.43	8.6	13.7
2×8	1.5	7.25	3.08	3.99	5.80	7.25	7.98	9.79	11.3	18.0
2×10	1.5	9.25	3.93	5.09	7.40	9.25	10.2	12.5	14.4	23.0
2×12	1.5	11.25	4.78	6.19	9.00	11.25	12.4	15.2	17.5	28.0
4×4	3.5	3.5	3.47	4.49	6.53	8.17	9.0	11.0	12.7	20.3
5×5	4.5	4.5	5.74	7.43	10.8	13.5	14.9	18.2	21.0	33.5
6×6	5.5	5.5	8.57	11.1	16.1	20.2	22.2	27.2	31.4	50.1
8×8	7.5	7.5	15.9	20.6	30.0	37.5	41.3	50.6	58.3	93.2
10×10	9.5	9.5	25.6	33.1	48.1	60.2	66.2	81.2	93.6	150
12×12	11.5	11.5	37.5	48.5	70.5	88.2	97.0	119	137	219
14×14	13.5	13.5	51.6	66.8	97.2	122	134	164	189	302
16×16	15.5	15.5	68.1	88.1	128	160	176	216	249	398
18×18	17.5	17.5	86.8	112	163	204	225	276	317	507
20×20	19.5	19.5	108	139	203	254	279	342	394	630
22×22	21.5	21.5	131	169	247	308	339	416	479	766
24×24	23.5	23.5	156	202	295	368	405	497	572	915

Jonathan S. Price

Harry Parker suggests adding between 50 and 67 percent to the cross-sectional area, in his *Simplified Design of Roof Trusses for Architects and Builders*:[3]

All tension members formed of timber have their sections reduced at joints by the necessary cutting for bolts and framing. Therefore, in the design of timber tension members, the cross section must

Metric Units

	Bald Cypress	Spruce-Pine-Fir	Redwood	Douglas Fir–Larch (N)	Glued Laminated	Southern Pine	LVL	LVL
Tension Strength, T (N)								
				F_t (kN/m²)				
	2,930	3,792	5,516	6,895	7,584	9,308	10,721	17,133
2×4	6.62	8.56	12.5	15.6	17.1	21.0	24.2	38.7
2×6	10.4	13.5	19.6	24.5	26.9	33.0	38.0	60.8
2×8	13.7	17.7	25.8	32.2	35.5	43.5	50.1	80.1
2×10	17.5	22.6	32.9	41.1	45.3	55.5	64.0	102
2×12	21.3	27.5	40.0	50.0	55.0	67.6	77.8	124
4×4	15.4	20.0	29.1	36.3	40.0	49.0	56.5	90.3
5×5	25.5	33.0	48.0	60.1	66.1	81.1	93.4	149
6×6	38.1	49.3	71.8	89.7	98.7	121	139	223
8×8	70.9	91.7	133	167	183	225	259	415
10×10	114	147	214	268	294	361	416	665
12×12	167	216	314	392	431	529	610	975
14×14	230	297	432	540	595	730	840	1,343
16×16	303	392	570	712	784	962	1,108	1,770
18×18	386	499	727	908	999	1,226	1,412	2,257
20×20	479	620	902	1,128	1,240	1,522	1,753	2,802
22×22	583	754	1,097	1,371	1,508	1,851	2,132	3,406
24×24	696	901	1,310	1,638	1,801	2,211	2,547	4,070

Notes: (1.) Apply appropriate adjustment factors; (2.) table values assume a 33% reduction in strength to account for connections and heavy timber reference design values

have a gross area in excess of the required **net area**. In timber trusses, it is customary to use timber for the lower chord, which resists tensile forces, the gross section being one-half to two-thirds greater than the net area theoretically required.

Following this recommendation, Table 3.1 provides tension strength for varying reference design stresses. This can be used to quickly find the initial member size for a given tension load.

Design is an iterative process, and the designer should verify initial selections after the details have been developed. Engineers need a system for differentiating preliminary sizes from members that have been fully designed. The author uses pencil for the preliminary design and electronic drawings for design development. Find something that works for you.

where

T = tension force, lb (kN)

A_{net} = is the net cross-sectional area, in^2 (m^2)

When the actual stress is less than the adjusted design stress, F'_t, we have a safe design.

In some cases, specifically for investigations of existing structures, the member sizes are known, but structural capacity is unknown. When we want to know the allowable tension value, T, we rearrange equation (3.5) so that T is on the left, and replace tension stress with the adjusted design stress, as follows:

$$T = F'_t A_{net} \tag{3.6}$$

This equation is the basis of the calculations in Table 3.1.

3.4 DEFLECTION

Pure tension deformation is illustrated in Figure 3.4. When we apply a tension force to our foam member, the circles deform and become elliptical—longer in the direction of tension.

Tension **deflection** tends to be rather small in members. However, in trusses, the tension and compression deflections add up to give deflections that we need to consider.

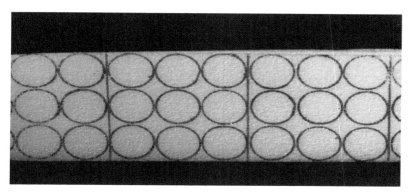

Figure 3.4 Tension deformation in a foam member

We calculate tension deflection as follows:

$$\delta = \frac{Tl}{AE'} \tag{3.7}$$

where:

T = axial tension force, lb (kN)

l = length, in (m)

A = cross-section area, in^2 (m^2), use gross area

E' = adjusted modulus of elasticity, lb/in^2 (kN/m^2)

Remember to watch your units. Also, refer to Chapter 4 for a discussion on long- and short-term deflection.

3.5 DETAILING

Connections are the biggest challenge in structural design. In historic structures, bottom chord and king-post connections were often dovetailed in, rather than the dubious tension capacity of mortise and tenon joints. Sometimes, the builders haunched the ends to avoid a reduction in cross-sectional area. After the industrial revolution, bolts and iron straps were often used to supplement truss tension connections. The tensile capacity of timber at connections became another required check.

In practice, after timber sizes have been selected, and the connections are designed, we need to verify that the timber net sections are adequate. In prefabricated trusses, such as the one shown in Figure 3.5, this responsibility falls on the truss manufacturer's engineer. Delegating the design responsibility is a mixed bag. The engineer of record (or the architect) retains veto authority if the connection does not fulfill the requirements.

Another tension connection type is the stud wall tie-down shown in Figure 3.6. Here, the tension force is in the studs, owing to lateral loads. The tension connection is made with a hold down attached to the studs and an anchor rod embedded in the foundation wall.

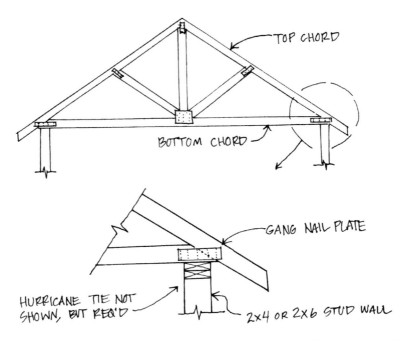

Figure 3.5 Prefabricated truss showing layout and bottom chord connection detail

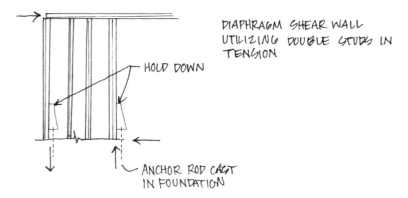

Figure 3.6 Stud wall tie-down

3.6 DESIGN STEPS

In general, the design of structural elements follow these steps:

1. Draw the structural layout (building geometry, column grid, structural member spans, and spacing).
2. Based on use and geographic location, determine the loads.
   (a.) **Live** and environmental loads are based on the building code.
   (b.) Building occupancy dictates code required live load.
   (c.) Refer to the building code for how to combine basic loading. The *IBC* specifies the minimum number of combinations that should be examined. One obvious gravity **load combination** is 1.0 Dead + 1.0 Live.
3. Material Parameters—Select the wood grade and species, and find the reference design values and adjustment factors.
4. Estimate initial size based on rules of thumb, preliminary tables, or a guess.
5. Calculate member stress and compare with adjusted reference stress.
6. Calculate deflections (usually not a factor for tension member design).
7. Summarize the results.

3.7 DESIGN EXAMPLE

Steps 1 and 2: Determine Layout and Loads

Assume you are designing a truss bottom chord and steps 1 and 2 above are complete, yielding the following bottom chord tension force:

$T = D + L$

$= 25\ k + 25\ k$	$= 111\ kN + 111\ kN$
$= 50\ k$	$= 222\ kN$

Note: The live load duration could be as long as 10 years.

Step 3: Determine Material Parameters

Assume you have decided to use Bald Cypress, No. 2. From Table A2.2:

$F_t = 425\ lb/in^2$	$F_t = 2{,}930\ kN/m^2$
$E = 1{,}000{,}000\ lb/in^2$	$E = 6{,}894{,}756\ kN/m^2$

The applicable adjustment factors are:

Factor	Description	Source
$C_D = 1.0$	Load duration—Live load	Table A4.1
$C_M = 1.0$	Wet service—Dry, interior condition	Table A4.2
$C_t = 1.0$	Temperature—Sustained temperatures don't exceed 100°F (37.8°C)	Table A4.4
$C_F = 1.0$	Size—Sawn lumber, guessing a 12 in (300 mm) deep member	Table A4.5
$C_i = 1.0$	Incising—Not treated wood	Table A4.10

Applying these to the reference design values, we get:

$F'_t = F_t C_D C_M C_t C_F C_i$

$= 425\ lb/in^2\ (1.0)$	$= 2{,}930\ kN/m^2\ (1.0)$
$= 425\ lb/in^2$	$= 2{,}930\ kN/m^2$

$E' = E_t C_M C_t C_i$

$= 1{,}000{,}000\ lb/in^2\ (1.0)$	$= 6{,}894{,}756\ kN/m^2\ (1.0)$
$= 1{,}000{,}000\ lb/in^2$	$= 6{,}894{,}756\ kN/m^2$

Step 4: Estimate Initial Size

Based on experience, we will try a 12 in (305 mm) square member.

Step 5: Calculate the Member Stresses

Rather than calculate member stress, let's calculate the required member size. Dividing the tension force by the adjusted design stress, F'_t, we get:

$$A_{net} = \frac{T}{F'_t}$$

$$= \frac{50,000 \text{ lb}}{425 \text{ lb/in}^2} \qquad\qquad = \frac{222 \text{ kN}}{2,930 \text{ kN/m}^2 \left(\dfrac{1 \text{ m}}{1000 \text{ mm}}\right)^2}$$

$$= 118 \text{ in}^2 \qquad\qquad\qquad = 78,768 \text{ mm}^2$$

Comparing this to the size of the trial member, from Table A1.1, we see that:

$$A = 132.3 \text{ in}^2 \qquad\qquad\qquad A = 0.085 \text{ m}^2$$

Assuming connections will decrease the net area, let's increase the size to 12 × 14 in (300 × 355 mm).

Step 6: Calculate Deflection

As a check on the stretching or elongation caused by tension, we use equation (3.7).

Suppose the bottom chord is 30 ft (9.14 m) long: the bottom chord stretch will be:

$$\delta = Tl/AE'$$

$$= \frac{50,000 \text{ lb } (30 \text{ ft}) \, 12 \text{ in/ft}}{11.5 \text{ in } (13.5 \text{ in}) \, 1,000,000 \text{ lb/in}^2} \qquad = \frac{222 \text{ kN } (9,140 \text{ mm})}{0.292 \text{ m } (0.340 \text{ m}) \, 6,894,756 \text{ kN/m}^2}$$

$$= 0.116 \text{ in} \qquad\qquad\qquad\qquad = 2.96 \text{ mm}$$

which is negligible.

Step 7: Summarize Results

We have selected a Bald Cypress No. 2 bottom chord that is 12 × 14 in (300 × 355 mm).

3.8 WHERE WE GO FROM HERE

Defining loads and determining a structural arrangement are fundamental for an efficient structural system. Before one launches into the design of individual members for tension, bending, or compression, the general arrangement of framing should be vetted by the architect or an experienced engineer.

Efficient framing systems acknowledge the architect's vision, while permitting some order and repetition. After the framing material has been selected—(timber for this book)—the designer usually performs some rudimentary calculations for gravity loads. Lateral loads should be considered at an early stage in the design, and bracing or shear walls should be located in a logical position to keep short and continuous load paths. Tension design of timber bracing, truss members, and, in some rare cases, tension hangers then follows the more difficult design tasks of building layout and global analysis noted above.

NOTES

1. Trautwine. *Civil Engineer's Reference Book* (1937).
2. ANSI/AWC. *National Design Specification NDS-2015* and the NDS Supplement—*Design Values for Wood Construction* (Leesburg, VA: AWC, 2015).
3. H. Parker. *Simplified Design of Roof Trusses for Architects and Builders* (New York: John Wiley, 1941).

Timber Bending

Chapter 4

Paul W. McMullin

Timber beams are used widely in all types of construction, ranging from simple homes to elaborate cathedrals. They are light, easy to modify on site, and economical—providing safe, simple shelter.

In this chapter, we will explore sawn, glued laminated, and engineered lumber beams and joists. We will learn the parameters to consider, some preliminary sizing tools, and how to do the in-depth calculations. This will prepare you to design beams, such as those in a new apartment (Figure 4.1), or evaluate them in a historic structure (Figure 4.2).

Timber beams and joists are made from either sawn lumber or engineered wood products, such as those in Figure 4.3. They can easily **span** distances from 10 to 50 ft (3–15 m), or further when needed. Floor and roof joists, spanning 10–25 ft (3–7.5 m), are commonly 6–12 in (150–300 mm) deep.

Before we go into the details, let's visualize how beams deform. Simple span beams with downward load experience tension in the bottom and compression in the top at the mid-span. Figure 4.4a shows a simply supported foam beam with a **point** load in the middle. Notice how the circles stretch in the bottom middle—indicating tension—and shorten at

Figure 4.1 I-joists and LVL beam in new multi-family construction

Paul W. McMullin

Figure 4.2 Timber beam in historic barn, Cane River Creole National Historical Park, Natchitoches, Louisiana

Source: Photo courtesy of Robert A. Young © 2007

Figure 4.3 Sawn, composite, and I-joist

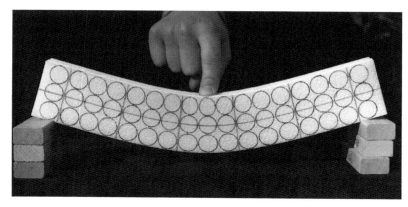

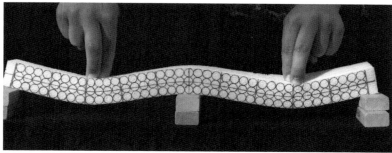

Figure 4.4 Foam beam showing deformation of (a) single and (b) double span

the top—indicating compression. Taking this further, Figure 4.4b shows a multi-span beam with point loads. The middle deformation is the same as for a simply supported beam, but, over the middle supports, where the beam is continuous, the tension and compression change places—tension on top, compression on bottom.

4.1 STABILITY

To be efficient, timber beams are deeper than they are wide. This moves more material away from the **neutral axis** (or center), increasing the moment of inertia, I, and section modulus, S. (Recall from Figure 2.31 that the depth is cubed and squared, respectively, for these properties, whereas the **width** is not.) However, deeper beams are prone to rolling over in the middle if they are not braced sufficiently.

Paul W. McMullin

To understand **lateral torsional buckling**, imagine you begin standing on the edge of a long, skinny board, braced at the ends. As you apply your weight, it begins to roll over to a flat position near the middle, shown in Figure 4.5. As this happens, the strength of the beam drops rapidly, and failure occurs. This is further illustrated in Figure 4.6, which shows the beam in a buckled state under the applied loads.

To keep this from happening, we need either to keep the beam proportions more like a square, or to brace it. We generally brace rectangular beams using one of the following methods:

- sheathing nailing—sufficient for joists;
- joists framing into a beam at regular intervals;
- beams framing into a beam;
- kickers bracing the bottom flange that is in compression over a support.

Stability considerations become more serious the deeper the beam gets relative to its width. We account for this with the beam stability factor, C_L. We consider three conditions:

- The beam is laid flatwise, or is square—$C_L = 1.0$.
- The beam is braced along its compression edge along its length— $C_L = 1.0$.
- The beam is not braced continuously—$C_L < 1.0$.

We calculate the beam stability factor utilizing the following equation:

$$C_L = \frac{1 + \left(F_{bE} \big/ F_b^* \right)}{1.9} - \sqrt{\left[\frac{1 + \left(F_{bE} \big/ F_b^* \right)}{1.9} \right]^2 - \frac{\left(F_{bE} \big/ F_b^* \right)}{0.95}}$$

(4.1)

where:

F_b^* = F_b multiplied by all the adjustment factors except C_{fu}, C_L, and C_V

$$F_{bE} = \frac{1.20 E'_{min}}{R_B^2}$$

(4.2)

E'_{min} = adjusted minimum elastic modulus, lb/in^2 (kN/m^2)

$$R_B = \sqrt{\frac{l_e d}{b^2}}$$

(4.3)

Figure 4.5 Lateral torsional buckling of a slender, unbraced beam showing (a–b) original condition and (c–f) increasing levels of buckling

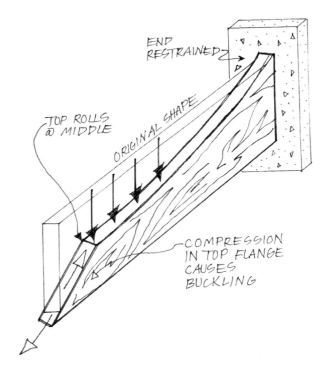

Figure 4.6 Lateral torsional buckling of floor joist

l_e = effective length, in (mm)—conservatively 2 times the **unbraced length**

d = depth, in (mm)

b = width, in (mm)

Getting through this analysis can be brutal. Take your time and you will get there. It helps to break the C_L equation into three numbers and then do the squaring and square rooting.

To illustrate how the beam stability factor affects the design, Figure 4.7 presents factors for different depth-to-width ratios and spans. Notice that, as the slenderness increases (larger d/b^2), the stability factor drops. Additionally, for longer beams, the stability factor also goes down for the same d/b^2 value.

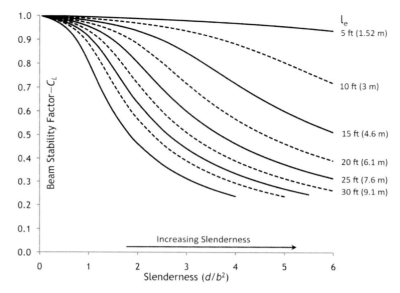

Figure 4.7 Beam stability factors, C_L, for varying spans and depths for Southern Pine No. 2 dense

4.2 CAPACITY

In timber design, we work in stress units, unlike for steel, concrete, and masonry, where we work in strength units. We first consult the reference design values—which give us our starting allowable stress or stiffness—and then the adjustment factors that modify these values for various influences. The strength (capacity) of timber design takes greater effort than the demand side.

4.2.1 Reference Design Values

Reference design values are at the core of timber design. The *NDS* tabulates these based on the following:

- species of wood;
- grade (quality);
- loading direction relative to the grain (i.e., parallel or perpendicular to grain, tension, compression, and bending).

Paul W. McMullin

With dozens of wood species, numerous grades for each, and a number of loading types, there is a multitude of timber reference design data, as seen in the *NDS Supplement*. Appendix 2 provides reference design values for a handful of species and grades. For bending design, we use allowable bending stress, F_b, and two moduli of elasticity, E and E_{min}—the former for deflection, the latter for stability.

To find the reference bending design stress, we begin by first deciding what wood product we will use, based on experience—sawn lumber, glued laminated timber, SCL, or I-joists. This will lead us to one of the following tables in Appendix 2

- visually graded dimension lumber—Table A2.1;
- visually graded timbers (5 × 5 in and larger)—Table A2.2;
- mechanically graded Douglas Fir–Larch (North)—Table A2.3;
- visually graded Southern Pine—Table A2.4;
- glued laminated timber—Table A2.5;
- SCL—Table A2.6;
- I-joists—Table A2.7.

If we are using sawn lumber, we additionally need a sense for species, size, and **grading** methods. These decisions are driven by local availability. (It is best if this is locally available and sustainably harvested.)

To find the reference design values, we enter one of the first six tables listed above. In the first column, select the species.

- Under the species heading, select the grade. (No. 2 is very common in Douglas Fir and Southern Pine. You may consider calling lumber suppliers in the project area to see what they commonly stock.)
- In the second column, choose the size class, if applicable.
- Finally, find the column that corresponds to the property you are looking for—F_b for bending in our case. Read down to the row for the species and grade you are using and read off the value in lb/in^2 (kN/m^2).

For example, say you want to know reference design stress, F_b, for Redwood, No. 1, open grain: you would enter Table A2.1 and scan down the first column until you found the line that says No. 1, open grain, under Redwood. You then scan to the right until you find the column with F_b in it. The number in this column and row combination is the reference bending design stress. It's not as hard as scaling a flaming volcano with only a calculator, but it takes some getting used to.

4.2.2 Adjusted Design Values

Now that we have the allowable reference design stress, we need to adjust it for the myriad variables we discussed in Chapter 2. Following Tables 2.4–2.7, we can quickly see which factors apply to bending for different wood products. For bending members, the equations for adjusted design stress are as follows:

$$F'_b = F_b C_D C_M C_t C_L C_F C_{fu} C_i C_r \text{ for sawn lumber} \tag{4.4}$$

$$F'_b = F_b C_D C_M C_t C_L C_V C_{fu} C_c C_I \text{ for glued laminated timber} \tag{4.5}$$

$$F'_b = F_b C_D C_M C_t C_L C_V C_r \text{ for SCL} \tag{4.6}$$

$$M'_r = M_r C_D C_M C_t C_r \text{ for I-joists} \tag{4.7}$$

where:

F_b = reference bending stress from Appendix 2

C_x = adjustment factors; see Table 2.3

Notice that, for I-joists, we work in capacity units of moment. This is because the manufacturers provide allowable moment, instead of section modulus and allowable stress. We then compare bending moment (demand) with adjusted bending design strength, M'_r (capacity). Also, we are not using the LRFD adjustment factors, as we are using the allowable stress method.

INITIAL BEAM SIZING

We frequently need to estimate beam depth early in a project, before we do calculations. Using simple rules of thumb based on span-to-depth ratios, we can estimate beam depth for a given span. A quick way to estimate joist size is to make the depth, in inches, equal to the span in feet divided by 2. For metric, multiply the span in meters by 40 to get the depth in millimeters. Multiply these by 1.5 for beams or **girders**.

Expanding on this, Table 4.1 provides beam depths for joists, sawn beams, glued laminated beams, and trusses based on span-to-depth ratios. To use the table, find your span along the top. Follow the column down to the row corresponding to your structural member type. Read off the depth in inches (millimeters).

Paul W. McMullin

Table 4.1 Beam member sizing guide

Imperial Units

System	Span/ Depth	Span (ft)								
		10	15	20	25	30	40	50	75	100
		Depth (in)								
Joist	22	6	8	12	14					
Solid Sawn Beam	16	8	12	16	20	24	30			
Glue–Lam Beam	20	6	9	12	15	18	24	30		
Truss	12	10	15	20	25	30	40	50	75	100

Metric Units

System	l/d	Span (m)								
		3	4.5	6	7.5	9	12	15	23	30
		Depth (mm)								
Joist	22	150	210	300	350					
Solid Sawn Beam	16	190	300	400	500	600	750			
Glue–Lam Beam	20	150	230	300	400	460	600	775		
Truss	12	150	380	500	630	760	1,010	1,300	1,900	2,500

DEPTH
SPAN

Table 4.2 Bending strength of sawn lumber for varying bending stresses

Imperial Units

	S_x in³	$F_b = 800\ lb/in^2$ C_L			$F_b = 1{,}500\ lb/in^2$ C_L			$F_b = 2{,}500\ lb/in^2$ C_L		
		1.0	0.7	0.4	1.0	0.7	0.4	1.0	0.7	0.4
2 × 4	3.06	0.20	0.14	0.08	0.38	0.27	0.15	0.64	0.45	0.26
2 × 6	7.56	0.50	0.35	0.20	0.95	0.66	0.38	1.58	1.10	0.63
2 × 8	13.14	0.88	0.61	0.35	1.64	1.15	0.66	2.74	1.92	1.10
2 × 10	21.39	1.43	1.00	0.57	2.67	1.87	1.07	4.46	3.12	1.78
2 × 12	31.64	2.11	1.48	0.84	3.96	2.77	1.58	6.59	4.61	2.64
6 × 12	121.2	8.08	5.66	3.23	15.2	10.61	6.06	25.3	17.7	10.10
6 × 16	220.2	14.7	10.3	5.87	27.5	19.3	11.01	45.9	32.1	18.35
6 × 20	348.6	23.2	16.3	9.30	43.6	30.5	17.43	72.6	50.8	29.05
6 × 24	506.2	33.7	23.6	13.5	63.3	44.3	25.3	105	73.8	42.2
10 × 16	380.4	25.4	17.8	10.1	47.5	33.3	19.0	79.2	55.5	31.7
10 × 20	602.1	40.1	28.1	16.1	75.3	52.7	30.1	125	87.8	50.2
10 × 24	874.4	58.3	40.8	23.3	109	76.5	43.7	182	128	72.9
14 × 18	689.1	45.9	32.2	18.4	86.1	60.3	34.5	144	100	57.4
14 × 24	1,243	82.8	58.0	33.1	155	108.7	62.1	259	181	104

Bending Strength, M (k-ft)

Metric Units

	$S_x \times 10^6$ mm³	Bending Strength M (kN-m)								
		$F_b = 5,500 \ kN/m^2$			$F_b = 10,350 \ kN/m^2$			$F_b = 17,200 \ kN/m^2$		
		C_L			C_L			C_L		
		1.0	0.7	0.4	1.0	0.7	0.4	1.0	0.7	0.4
2 × 4	0.050	0.28	0.19	0.11	0.52	0.36	0.21	0.86	0.61	0.35
2 × 6	0.124	0.68	0.48	0.27	1.28	0.90	0.51	2.14	1.49	0.85
2 × 8	0.215	1.19	0.83	0.48	2.23	1.56	0.89	3.71	2.60	1.48
2 × 10	0.351	1.93	1.35	0.77	3.63	2.54	1.45	6.04	4.23	2.42
2 × 12	0.518	2.86	2.00	1.14	5.36	3.75	2.14	8.94	6.26	3.57
6 × 12	1.987	11.0	7.67	4.38	20.5	14.38	8.22	34.2	24.0	13.70
6 × 16	3.609	19.9	13.9	7.96	37.3	26.1	14.93	62.2	43.5	24.88
6 × 20	5.712	31.5	22.1	12.60	59.1	41.4	23.63	98.5	68.9	39.38
6 × 24	8.296	45.8	32.0	18.3	85.8	60.1	34.3	143	100	57.2
10 × 16	6.234	34.4	24.1	13.8	64.5	45.1	25.8	107	75.2	43.0
10 × 20	9.866	54.4	38.1	21.8	102	71.4	40.8	170	119	68.0
10 × 24	14.33	79.0	55.3	31.6	148	104	59.3	247	173	98.8
14 × 18	11.29	62.3	43.6	24.9	117	81.7	46.7	195	136	77.9
1 4 × 24	20.36	112	78.6	44.9	211	147.4	84.2	351	246	140

Notes: (1.) This table is for preliminary sizing only. Final section sizes must be calculated based on actual loading, length, and section size. (2.) Span ranges indicated are typical. Longer spans can be made with special consideration

Taking things further, Tables 4.2 and 4.3 provide bending strength for varying section sizes, timber reference stresses, and beam stability factors. Table A2.7 provides these for I-joists. We can

Table 4.3 **Bending strength of engineered lumber for various products**

Imperial Units		*Bending Strength M (k-ft)*					
		C_L			C_L		
	S_x in^3	**1.0**	0.7	0.4	**1.0**	0.7	0.4
Laminated Veneer Lumber (LVL)							
		$F_b = 2,140 \ lb/in^2$			$F_b = 2,900 \ lb/in^2$		
$1^3/_4 \times 5^1/_2$	8.82	**1.57**	1.10	0.63	**2.13**	1.49	0.85
$1^3/_4 \times 7^1/_4$	15.3	**2.73**	1.91	1.09	**3.70**	2.59	1.48
$1^3/_4 \times 9^1/_2$	26.3	**4.69**	3.29	1.88	**6.36**	4.45	2.54
$1^3/_4 \times 11^7/_8$	41.1	**7.33**	5.13	2.93	**9.94**	6.96	3.98
$1^3/_4 \times 14$	57.2	**10.2**	7.14	4.08	**13.8**	9.67	5.53
$1^3/_4 \times 16$	74.7	**13.3**	9.32	5.33	**18.0**	12.6	7.22
$1^3/_4 \times 18$	94.5	**16.9**	11.8	6.74	**22.8**	16.0	9.14
Glued Laminated Timber							
		$F_b = 1,600 \ lb/in^2$			$F_b = 2,400 \ lb/in^2$		
$3^1/_8 \times 12$	75.0	**10.0**	7.00	4.00	**15.0**	10.5	6.00
$3^1/_8 \times 18$	168.8	**22.5**	15.8	9.0	**33.8**	23.6	13.5
$3^1/_8 \times 24$	300.0	**40.0**	28.0	16.0	**60.0**	42.0	24.0
$5^1/_8 \times 18$	276.8	**36.9**	25.8	14.8	**55.4**	38.7	22.1
$5^1/_8 \times 21$	376.7	**50.2**	35.2	20.1	**75.3**	52.7	30.1
$5^1/_8 \times 27$	622.7	**83.0**	58.1	33.2	**125**	87.2	49.8
$5^1/_8 \times 33$	930.2	**124**	86.8	49.6	**186**	130	74.4
$10^3/_4 \times 36$	2,322	**310**	217	124	**464**	325	186
$10^3/_4 \times 48$	4,128	**550**	385	220	**826**	578	330
$10^3/_4 \times 60$	6,450	**860**	602	344	**1,290**	903	516

Paul W. McMullin

compare calculated moment to those in the tables to get a good sense for initial size. From there, we can run the full calculations for our specific conditions. Remember to consider how the strength adjustment factors may affect your initial size.

Metric Units							
			Bending Strength M (kN-m)				
	$S_x \times 10^6$	C_L			C_L		
	mm^3	*1.0*	*0.7*	*0.4*	*1.0*	*0.7*	*0.4*
Laminated Veneer Lumber (LVL)							
		$F_b = 14{,}750\ kN/m^2$			$F_b = 20{,}000\ kN/m^2$		
$1^3/_4 \times 5^1/_2$	0.145	**2.13**	1.49	0.85	**2.89**	2.02	1.16
$1^3/_4 \times 7^1/_4$	0.251	**3.71**	2.59	1.48	**5.02**	3.52	2.01
$1^3/_4 \times 9^1/_2$	0.431	**6.36**	4.46	2.55	**8.62**	6.04	3.45
$1^3/_4 \times 11^7/_8$	0.674	**9.94**	6.96	3.98	**13.48**	9.43	5.39
$1^3/_4 \times 14$	0.937	**13.8**	9.68	5.53	**18.7**	13.11	7.49
$1^3/_4 \times 16$	1.22	**18.1**	12.64	7.22	**24.5**	17.1	9.79
$1^3/_4 \times 18$	1.55	**22.8**	16.0	9.14	**31.0**	21.7	12.39
Glued Laminated Timber							
		$F_b = 11{,}030\ kN/m^2$			$F_b = 16{,}550\ kN/m^2$		
$3^1/_8 \times 12$	1.23	**13.6**	9.49	5.42	**20.3**	14.2	8.13
$3^1/_8 \times 18$	2.77	**30.5**	21.4	12.2	**45.8**	32.0	18.3
$3^1/_8 \times 24$	4.92	**54.2**	38.0	21.7	**81**	56.9	32.5
$5^1/_8 \times 18$	4.54	**50.0**	35.0	20.0	**75.0**	52.5	30.0
$5^1/_8 \times 21$	6.2	**68.1**	47.7	27.2	**102**	71.5	40.9
$5^1/_8 \times 27$	10.2	**113**	78.8	45.0	**169**	118	68
$5^1/_8 \times 33$	15.2	**168**	118	67.3	**252**	177	101
$10^3/_4 \times 36$	38.1	**420**	294	168	**630**	441	252
$10^3/_4 \times 48$	67.6	**746**	522	298	**1,119**	784	448
$10^3/_4 \times 60$	106	**1,166**	816	466	**1,749**	1,224	700

4.3 DEMAND VS. CAPACITY

Once we know the adjusted design stress, F'_b, we compare it with the bending stress. For a simply supported beam, stress varies triangularly from compression at the top to tension at the bottom—zero at the middle—as illustrated in Figure 4.8a. A cantilever beam flips the stress direction upside down—tension at the top near the support, compression at the bottom—shown in Figure 4.8b. We also know bending stress changes along the beam length, as illustrated in Figure 4.9 for several support conditions. In the single-span, simply supported beam case, the bending stress is zero at the ends and maximum at the middle. A cantilever is the opposite, with maximum stress at the supported end. A multi-span beam has positive bending stress at the mid-spans, but negative bending stress over the interior supports. Positive means tension bending stress at the bottom; negative indicates tension stress at the top.

Regardless of the moment distribution along the beam length, we find bending stress from equation (4.8)—paying attention to our units. Though we are typically concerned with the maximum moment (and therefore stress), we can insert the moment at any point along the beam and find its corresponding bending stress.

$$f_b = \frac{M}{S} \quad \text{or} \quad f_b = \frac{Mc}{I} \tag{4.8}$$

where:

M = bending moment at location of interest, k-ft (kN-m)

S = section modulus, in^3 (mm^3)

c = distance from neutral axis to compression or tension face, in (mm)

I = moment of inertia, in^4 (mm^4)

As long as the bending stress is less than the adjusted design stress, we are OK. If it is higher, we select a larger beam and re-evaluate the bending stresses before moving into shear and deflection checks.

4.3.1 Sheathing Thickness

Sizing sheathing is quite different than beams and joists. Panel strength is determined by span ratings. These are either combined roof/floor ratings or single-floor grade spans. A combined roof/floor rating gives the roof span followed by the floor span—for example, 48/24 indicates the panel

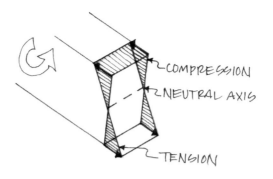

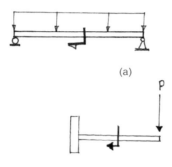

(a)

BENDING MOMENT

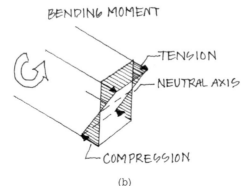

(b)

Figure 4.8 Bending distribution stress in (a) simple supported beam, (b) cantilever beam

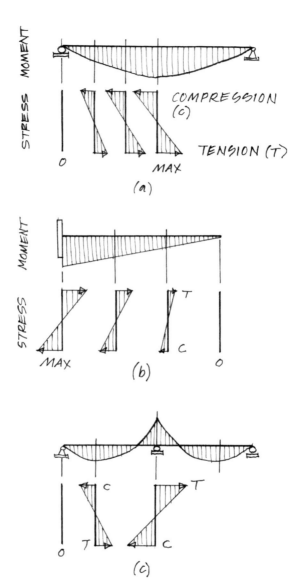

Figure 4.9 Bending stress variation along beam length for (a) single-span, simple support, (b) cantilever, (c) multi-span

Paul W. McMullin

can span 48 in (1,220 mm) on a roof and 24 in (610 mm) on a floor. Single-floor ratings provide the allowable panel span for a floor (e.g., 24 on center—o.c.), though it can still be used on roofs.

To select a roof panel thickness, enter Table 4.4 with your structural member spacing and the total load and live load you need to support. Find the maximum span you have and verify that the loads you have are less than the allowables. Then, read to the left-most two columns and get the span rating and panel thickness. For floors, find the span you need in the far-right column, and read the span rating and thickness from the far-left columns.

Table 4.4 Allowable spans and loads for structural panel sheathing

	Imperial Units		*Roof*				*Floor*
	Panel		*Max Span (in)*		*Load (lb/ft²)*		*Span (in)*
	Rating	*Thickness (in)*	*Edge Support*	*No Edge Support*	*Total*	*Live*	
Roof/Floor	16/0	³/₈	16	16	40	30	0
	20/0	³/₈	20	20	40	30	0
	24/0	³/₈, ⁷/₁₆, ¹/₂	24	20	40	30	0
	24/16	⁷/₁₆, ¹/₂	24	24	50	40	16
	32/16	¹⁵/₃₂, ¹/₂, ⁵/₈	32	28	40	30	16
	40/20	¹⁹/₃₂, ⁵/₈, ³/₄, ⁷/₈	40	32	40	30	20
	48/24	²³/₃₂, ³/₄, ⁷/₈	48	36	45	35	24
	54/32	⁷/₈, 1	54	40	45	35	32
	60/32	⁷/₈, 1¹/₈	60	48	45	35	32
Single Floor	16 o.c.	¹/₂, ¹⁹/₃₂, ⁵/₈	24	24	50	40	16
	20 o.c.	¹⁹/₃₂, ⁵/₈, ³/₄	32	32	40	30	20
	24 o.c.	²³/₃₂, ³/₄	48	36	35	25	24
	32 o.c.	⁷/₈, 1	48	40	50	40	32
	48 o.c.	1³/₃₂, 1¹/₈	60	48	50	40	48

Note: (1.) Allowable floor load is 100 lb/ft² except for 48 o.c., which is 65 lb/ft²

Table 4.4 *continued*

Metric Units			Roof			Floor
Panel			*Max Span (mm)*		*Load (kN/m²)*	*Span (mm)*
Rating	*Thickness (mm)*		*Edge Support*	*No Edge Support*	*Total*	*Live*	
Roof/Floor 16/0	9.5		400	400	1.92	1.44	0
20/0	9.5		500	500	1.92	1.44	0
24/0	9.5, 11, 13		600	500	1.92	1.44	0
24/16	11, 13		600	600	2.39	1.92	400
32/16	12, 13, 16		810	710	1.92	1.44	400
40/20	15, 16, 19, 22		1,010	810	1.92	1.44	500
48/24	18, 19, 22		1,210	910	2.15	1.68	600
54/32	22, 25		1,370	1,010	2.15	1.68	800
60/32	22, 29		1,520	1,210	2.15	1.68	800
Single Floor 16 o.c.	13, 15, 16		600	600	2.39	1.92	400
20 o.c.	15, 16, 19		810	810	1.92	1.44	500
24 o.c.	18, 19		1,210	910	1.68	1.20	600
32 o.c.	22, 25		1,210	1,010	2.39	1.92	800
48 o.c.	28, 29		1,520	1,210	2.39	1.92	1,200

Note: (1.) Allowable floor load is 4.79 kN/m² except for 48 o.c., which is 3.11 kN/m²
Source: IBC 2012

4.4 DEFLECTION

In addition to stability and strength we need to look at how much a beam will deflect (sag), known as serviceability criteria. Excessive deflection can cause floor and ceiling finishes to crack, and windows or doors to carry load, binding or cracking them.

Under sustained loads, timber deflection increases over time. This is known as **creep**. The barn in Figure 4.10 well illustrates the creep phenomenon. Creep also happens to masonry and concrete at room temperature, and to steels at high temperatures. In timber beams, creep

Paul W. McMullin

causes us to look at short- and **long-term deflections**, yielding a total deflection equation of:

$$\delta_T = K_{cr}\delta_{LT} + \delta_{ST} \qquad (4.9)$$

where:

K_{cr} = creep deformation factor
    = 1.5 for materials in dry service (except as noted below)
    = 2.0 for materials in wet service
    = 2.0 for unseasoned lumber, structural panels, cross-laminated timber

δ_{LT} = long-term deflection, in (mm)

δ_{ST} = **short-term deflection**, in (mm)

We calculate both long- and short-term deflections using the structural analysis equations in Appendix 6, or similar sources. For long-term calculations, we use any permanent loads (typically dead); for short-term calculations, we use any transient loads (live, **snow**, wind, seismic). We always use allowable stress combinations, as we are checking an in-service condition, not a strength limit state.

Figure 4.10 Old barn showing creep deformation
Source: Photo courtesy of muenstermann

We compare deflections to the code allowables provided in Table 4.5. Expanding on these, Table 4.6 calculates allowable deflection, δ_a, values for various spans and code limits. If our total and live load deflections are less than the allowables, our beam is stiff enough. If not, we choose a larger member, or reduce the load on it.

4.5 DETAILING

Timber beam detailing centers more around the connections than the members themselves—discussed in Chapter 9. However, it is worth discussing stability bracing, notches, and holes.

Stability bracing is required for beams, joists, and rafters to keep them from rolling over. Ends of members must also be braced against rolling. Along the length, we brace the member either with sheathing or other members—unless we account for the lack of bracing in the calculations

Table 4.5 Deflection limits for beams

Deflection Limits		
Member	*L or S or W*	*D + L*
Roof members:		
Plaster or stucco ceiling	1/360	1/240
Non-plaster ceiling	1/240	1/180
No ceiling	1/180	1/120
Floor members:		
Typical	1/360	1/240
Tile	1/480	1/360
Supporting masonry	1/600	1/480
Wall members:		
Plaster or stucco	1/360	–
Other brittle finishes	1/240	–
Flexible finishes	1/120	–

Note: l = span; don't forget to convert to inches
Source: IBC 2012

Paul W. McMullin

Table 4.6 Calculated allowable deflection values for various spans and limits

Allowable Deflection Values for Various Criteria & Lengths

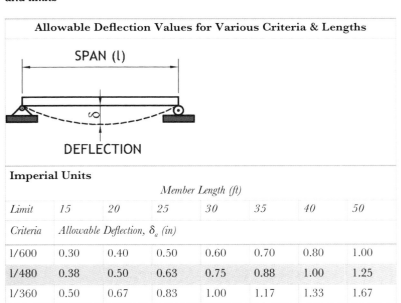

Imperial Units

Member Length (ft)

Limit	15	20	25	30	35	40	50
Criteria	*Allowable Deflection, δ_a (in)*						
1/600	0.30	0.40	0.50	0.60	0.70	0.80	1.00
1/480	0.38	0.50	0.63	0.75	0.88	1.00	1.25
1/360	0.50	0.67	0.83	1.00	1.17	1.33	1.67
1/240	0.75	1.00	1.25	1.50	1.75	2.00	2.50
1/180	1.00	1.33	1.67	2.00	2.33	2.67	3.33
1/120	1.50	2.00	2.50	3.00	3.50	4.00	5.00

Metric Units

Member Length (m)

Limit	4	6	8	9	10	12	15
Criteria	*Allowable Deflection, δ_a (mm)*						
1/600	6.7	10	13	15	17	20	25
1/480	8.3	13	17	19	21	25	31
1/360	11	17	22	25	28	33	42
1/240	17	25	33	38	42	50	63
1/180	22	33	44	50	56	67	83
1/120	33	50	67	75	83	100	125

(C_L). For joists, we add **bridging** or blocking every 8 ft (2.4 m) if $d/b > 6$, as illustrated in Figure 4.11, to stabilize the floor system.

From time to time, the contractor will notch or drill holes in beams and joists. Naturally, this is not preferred, but it is the reality of construction. In sawn beams, the *NDS* limits the depth and locations of notches in the end third of a span (and top through the middle), according to the requirements of Figure 4.12. No bottom notches are permitted through the center of the span. Further, in floor and roof joists, the code permits small holes to accommodate piping and electrical wiring. These holes must be 2 in (50 mm) away from the edges, no larger than ⅓ of the depth, as illustrated in Figure 4.13, and spaced reasonably apart. Engineered-lumber manufacturers allow some notches and holes, but they vary substantially according to the product type. Before cutting notches or drilling holes in members, check with your structural engineer. He or she will want to know, and may have good reason not to permit them.

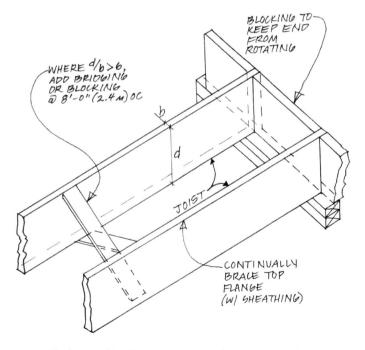

Figure 4.11 Blocking and bridging requirements for joists and rafters

Paul W. McMullin

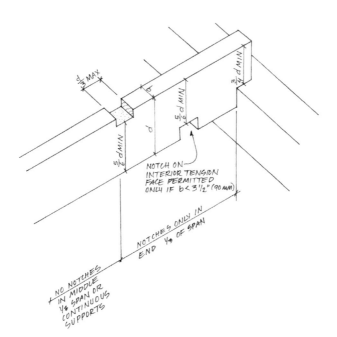

Figure 4.12 Permissible notches in sawn lumber

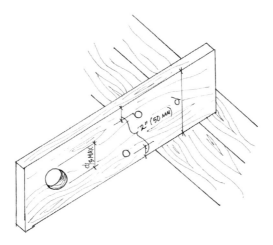

Figure 4.13 Permissible hole locations in sawn joists and rafters

Timber Bending

4.6 DESIGN STEPS

1. Draw the structural layout; include span dimensions and tributary width.
2. Determine loads:
   (a) unit loads;
   (b) load combinations yielding a line load;
   (c) member moment.
3. Material parameters: Find the reference design values and adjustment factors.
4. Estimate initial size.
5. Calculate stress and compare with adjusted design stress.
6. Calculate deflection and allowable deflections, and compare them.
7. Summarize the results.

4.7 DESIGN EXAMPLES

4.7.1 Beam Example

This example sizes a sawn lumber and glued laminated timber beam for the framing layout shown in Figure 4.14. We will design the joists in the next examples.

Step 1: Draw Structural Layout

We begin by drawing the framing layout, complete with the dimensions we will need in the design (Figure 4.14). Key dimensional data include length, l, and tributary width, l_t:

l = 20 ft $\qquad\qquad$ l = 6.10 m

l_t = 16 ft $\qquad\qquad$ l_t = 4.88 m

Step 2: Determine Loads

Step 2a: Unit Loads

The unit dead and live loads are:

$q_D = 20 \dfrac{\text{lb}}{\text{ft}^2}$ $\qquad\qquad$ $q_D = 0.958 \dfrac{\text{kN}}{\text{m}^2}$

$q_L = 50 \dfrac{\text{lb}}{\text{ft}^2}$ $\qquad\qquad$ $q_L = 2.394 \dfrac{\text{kN}}{\text{m}^2}$

92 $\qquad\qquad\qquad$ Paul W. McMullin

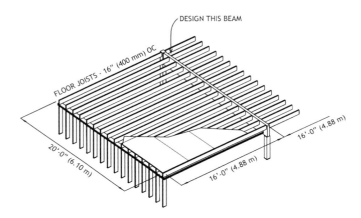

Figure 4.14 Example framing configuration

See Chapter 8 of *Introduction to Structures* for a discussion on how to determine these unit loads.

Step 2b: Load Combination

Because this is a floor, the live load dominant combination will control. Multiplying it by the tributary width, we find the line load, *w*, as:

$$w = (q_D + q_L)l_t$$

$$= \left(20\frac{\text{lb}}{\text{ft}^2} + 50\frac{\text{lb}}{\text{ft}^2}\right)16\text{ ft} \qquad = \left(0.958\frac{\text{kN}}{\text{m}^2} + 2.394\frac{\text{kN}}{\text{m}^2}\right)4.88\text{ m}$$

$$= 1{,}120\frac{\text{lb}}{\text{ft}} \qquad\qquad\qquad = 16.4\frac{\text{kN}}{\text{m}}$$

Step 2c: Determine Member Moment

We are only concerned with the maximum moment, which occurs at the middle. Using the formulas in Appendix 6 and Figure 4.15, we see:

$$M = \frac{wl^2}{8}$$

$$= \frac{1{,}120\text{ lb/ft }(20\text{ ft})^2}{8}\frac{1\text{ k}}{1000\text{ lb}} \qquad = \frac{16.4\text{ kN/m }(6.1\text{ m})^2}{8}$$

$$= 56.0\text{k} - \text{ft} \qquad\qquad\qquad = 76.3\text{ kN} - \text{m}$$

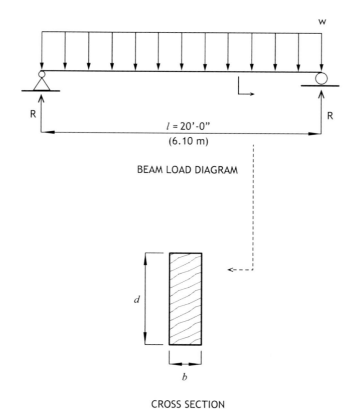

BEAM LOAD DIAGRAM

CROSS SECTION

Figure 4.15 Example beam free body diagram and cross section

Step 3: Material Parameters

We will look at both a sawn lumber beam and glued laminated beam. Using Tables A2.4 and A2.5, we will find the reference bending stress for Southern Pine. Note we are using subscript S for sawn lumber and G for glued laminated timber.

For sawn lumber, No. 1:

$$F_{bS} = 1,250 \frac{\text{lb}}{\text{in}^2}$$

$$F_{bS} = 8,618 \frac{\text{kN}}{\text{m}^2}$$

For a glued laminated 24F-V3 layup:

$$F_{bG} = 2,400 \frac{\text{lb}}{\text{in}^2} \qquad\qquad F_{bG} = 16,547 \frac{\text{kN}}{\text{m}^2}$$

We now apply the adjustment factors to these, following the list in Tables 2.4 and 2.5. We will use the factors shown in the following table:

Factor	Description	Source
$C_D = 1.0$	Load duration—floor load	Table A4.1
$C_M = 1.0$	Wet service—dry, interior condition	Table A4.2
$C_t = 1.0$	Temperature—sustained temperatures don't exceed 100°F (37.8°C)	Table A4.4
$C_L = 1.0$	Beam stability—beam is fully braced	Eqn. (4.1)
$C_F = 0.9$	Size—sawn lumber, guessing a 24 in (600 mm) deep member	Table A4.5
$C_V = 0.94$	Volume—glued laminated	Table A4.6
$C_{fu} = 1.0$	Flat use—not laid flatwise	Table A4.8
$C_i = 1.0$	Incising—not treated wood	Table A4.10
$C_r = 1.0$	Repetitive member—isolated	Section 2.4.11
$C_c = 1.0$	Curvature—not curved	Table A4.9
$C_I = 1.0$	Stress interaction—not tapered	Section 2.4.8

Multiplying these together with the reference design stress, we get the adjusted design stress.

For sawn lumber (S):

$$F'_{bS} = F_{bS} C_D C_M C_t C_L C_F C_{fu} C_i C_r$$

$$= 1,250 \frac{\text{lb}}{\text{in}^2} 1.0\,(0.9)\,1.0 = 1,125 \frac{\text{lb}}{\text{in}^2} \qquad = 8,618 \frac{\text{kN}}{\text{m}^2} 1.0\,(0.9)\,1.0 = 7,756 \frac{\text{kN}}{\text{m}^2}$$

For glued laminated timber (G):

$$F'_{bG} = F_{bG} C_D C_M C_t C_L C_V C_{fu} C_c C_I$$

$$= 2,400 \frac{\text{lb}}{\text{in}^2} 1.0 \ (0.94) \ 1.0 \qquad\qquad = 16,547 \frac{\text{kN}}{\text{m}^2} 1.0 \ (0.94) \ 1.0$$

$$= 2,256 \frac{\text{lb}}{\text{in}^2} \qquad\qquad\qquad\qquad = 15,554 \frac{\text{kN}}{\text{m}^2}$$

While we're at it, let's find the adjusted modulus of elasticity for sawn lumber. We will need it for the deflection calculation.

$$E = 1,600,000 \frac{\text{lb}}{\text{in}^2} \qquad\qquad\qquad E = 11,032 \frac{\text{MN}}{\text{m}^2}$$

$$E' = E C_M C_t C_i$$

$$= 1,600,000 \frac{\text{lb}}{\text{in}^2}(1.0) = 1,600,000 \frac{\text{lb}}{\text{in}^2} \qquad = 11,032 \frac{\text{MN}}{\text{m}^2}(1.0) = 11,032 \frac{\text{MN}}{\text{m}^2}$$

Step 4: Initial Size

Following the Initial Beam Sizing box, we will say the beam depth is equal to half the span in inches times 1.5 (or multiply the span in meters by 40 to get depth in millimeters).

$$d_{\text{est}} = \frac{l}{2} 1.5 \qquad\qquad\qquad d_{\text{est}} = 40l \ (1.5)$$

$$= \frac{20}{2} 1.5 = 15 \ \text{in} \qquad\qquad = 40 \ (6.1) \ (1.5) = 366 \ \text{mm}$$

Using the section properties table in Appendix 1 (Table A1.1), let's try a 16 in (400 mm) deep member and make the width half the depth, giving us:

$d = 15.5$ in $d = 400$ mm

$b = 7.5$ in $b = 190$ mm

Step 5: Stress

Now that we have our size, we can calculate the section modulus, or refer to Table A1.1.

$$S = \frac{1}{6} bd^2$$

$$= \frac{1}{6}(7.5 \ \text{in}) \ (15.5 \ \text{in})^2 \qquad\qquad = \frac{1}{6}(190 \ \text{mm}) \ (400 \ \text{mm})^2$$

$$= 300.3 \ \text{in}^2 \qquad\qquad\qquad = 5.07 \times 10^6 \ \text{mm}^3$$

Paul W. McMullin

We then calculate stress, adjusting the units:

$$f_b = \frac{M}{S}$$

$$= \frac{56.0 \text{ k} - \text{ft}}{300.3 \text{ in}^3} 12 \frac{\text{in}}{\text{ft}} 1000 \frac{\text{lb}}{\text{k}}$$

$$= 2{,}238 \frac{\text{lb}}{\text{in}^2}$$

$$= \frac{76.3 \text{ kN} - \text{m}}{5.07 \times 10^6 \text{ mm}^3 \left(1 \text{ m} \middle/ 1000 \text{ mm}\right)^3}$$

$$= 15{,}050 \frac{\text{kN}}{\text{m}^2}$$

Comparing this with the allowables for both sawn and glued laminated, we see we are close for glued laminated, but have double the stress for sawn. We can either use a higher-grade sawn material, or go larger. Let's try larger.

To get really close on sawn size, let's calculate the required section modulus directly.

$$S_{\text{req}} = \frac{M}{F'_{bS}}$$

$$= \frac{56.0 \text{ k} - \text{ft}}{1{,}125 \text{ lb/in}^2} 12 \frac{\text{in}}{\text{ft}} 1000 \frac{\text{lb}}{\text{k}}$$

$$= 597 \text{ in}^3$$

$$= \frac{76.3 \text{ kN} - \text{m}}{7{,}756 \text{ kN/m}^2} \left(\frac{1000 \text{ mm}}{1 \text{ m}}\right)^3$$

$$= 9.84 \times 10^6 \text{ mm}^3$$

Now, going directly to Table A1.1, we can find a size with a slightly larger section modulus. Let's try 8 × 24 (203 × 610 mm).

$$S = 690.3 \text{ in}^3 \qquad\qquad S = 11.35 \times 10^6 \text{ mm}^3$$

$$f_b = \frac{M}{S}$$

$$= \frac{56.0 \text{ k} - \text{ft}}{690.3 \text{ in}^3} 12 \frac{\text{in}}{\text{ft}} 1000 \frac{\text{lb}}{\text{k}}$$

$$= 974 \frac{\text{lb}}{\text{in}^2}$$

$$= \frac{76.3 \text{ kN} - \text{m}}{11.35 \times 10^6 \text{ mm}^3 \left(1 \text{ m} \middle/ 1000 \text{ mm}\right)^3}$$

$$= 6{,}723 \frac{\text{kN}}{\text{m}^2}$$

We see the new bending stress is below the adjusted design stress value for sawn lumber. Cool!

Now, let's refine the glued laminated timber calculation, as we haven't selected a final size. Knowing our initial size was close, we choose a beam with a similar section modulus.

$$d = 18 \text{ in} \qquad\qquad d = 457 \text{ mm}$$
$$b = 6.75 \text{ in} \qquad\qquad b = 171 \text{ mm}$$

$$S = \frac{1}{6}bd^2$$

$$= \frac{1}{6}6.75 \text{ in } (18 \text{ in})^2 \qquad\qquad = \frac{1}{6}171 \text{ mm } (457 \text{ mm})^2$$

$$= 364.5 \text{ in}^3 \qquad\qquad = 5.95 \times 10^6 \text{ mm}^3$$

This yields a bending stress in the glued laminated beam of

$$f_b = \frac{M}{S}$$

$$= \frac{56.0 \text{ k}-\text{ft}}{364.5 \text{ in}^3}12\frac{\text{in}}{\text{ft}}1000\frac{\text{lb}}{\text{k}} \qquad = \frac{76.3 \text{ kN}-\text{m}}{5.95 \times 10^6 \text{ mm}^3\left(1 \text{ m}/1000 \text{ mm}\right)^3}$$

$$= 1{,}844\frac{\text{lb}}{\text{in}^2} \qquad\qquad = 12{,}820 \frac{\text{kN}}{\text{m}^2}$$

This is less than F'_{bG} above, indicating our beam works for stress.

Step 6: Deflection

Now that we have sizes and know the stresses are low enough, we check deflection to make sure we don't have a beam that sags too much. We will just check the sawn beam. For enlightenment, you can check the glued laminated beam.

The general equation for deflection is

$$\delta_{\text{TOT}} = K_{\text{cr}}\delta_{\text{LT}} + \delta_{\text{ST}}$$

$$K_{\text{cr}} = 1.5$$

$$\delta_{LT} = \frac{5w_D l^4}{384E'I} \qquad\qquad \delta_{ST} = \frac{5w_L l^4}{384E'I}$$

We will need to pause and calculate a few of these terms:

$$w_D = q_D l_t$$

$$= 20\frac{\text{lb}}{\text{ft}^2}16 \text{ ft} = 320\frac{\text{lb}}{\text{ft}} \qquad = 0.958\frac{\text{kN}}{\text{m}^2}4.88 \text{ m} = 4.68\frac{\text{kN}}{\text{m}}$$

$$w_L = q_L l_t$$

$$= 50\frac{\text{lb}}{\text{ft}^2}16 \text{ ft} = 800\frac{\text{lb}}{\text{ft}} \qquad = 2.39 \frac{\text{kN}}{\text{m}^2}4.88 \text{ m} = 11.7\frac{\text{kN}}{\text{m}}$$

For the sawn timber:

$$I = 8{,}111 \text{ in}^4 \qquad\qquad I = 3.387 \times 10^9 \text{mm}^4 \text{ from Table A1.1}$$

With these, we can now calculate deflection. Starting with long-term (dead) deflection:

$$\delta_{LT} = \frac{5(320 \text{ lb/ft})(20 \text{ ft})^4}{384(1,600,000 \text{ lb/in}^2)\,8,111 \text{ in}^4}\left(12\frac{\text{in}}{\text{ft}}\right)^3$$

$$= 0.088 \text{ in}$$

$$\delta_{LT} = \frac{5(4.68 \text{ kN/m})(6.10 \text{ m})^4}{384(11,032 \text{ MN/m}^2)\,3.387\times10^9 \text{ mm}^4}\left(\frac{1,000 \text{ mm}/1 \text{ m}}{1,000 \text{ kN}/1 \text{ MN}}\right)^5$$

$$= 2.26 \text{ mm}$$

And now, short-term (live load) deflection:

$$\delta_{ST} = \frac{5(800 \text{ lb/ft})(20 \text{ ft})^4}{384(1,600,000 \text{ lb/in}^2)\,8,111 \text{ in}^4}\left(12\frac{\text{in}}{\text{ft}}\right)^3$$

$$= 0.222 \text{ in}$$

$$\delta_{LT} = \frac{5(11.7 \text{ kN/m})(6.10 \text{ m})^4}{384(11,032 \text{ MN/m}^2)\,3.387\times10^9 \text{ mm}^4}\left(\frac{1,000 \text{ mm}/1 \text{ m}}{1,000 \text{ kN}/1 \text{ MN}}\right)^5$$

$$= 5.65 \text{ mm}$$

$\delta_{TOT} = 1.5(0.088 \text{ in}) + 0.222 \text{ in}$ $\qquad$ $\delta_{TOT} = 1.5(2.2\,6 \text{ mm}) + 5.65 \text{ mm}$

$\quad = 0.354 \text{ in}$ $\qquad\qquad\qquad\qquad$ $= 9.04 \text{ mm}$

This is cool, but we need to compare it with a standard. Because this is a floor that may have a gypsum board ceiling, we need to minimize deflection. According to the building code, the total load deflection should be limited to:

$$\delta_{aTOT} = \frac{l}{240}$$

$= \dfrac{20 \text{ ft}}{240}\dfrac{12 \text{ in}}{1 \text{ ft}} = 1 \text{ in}$ $\qquad\qquad$ $= \dfrac{6.10 \text{ m}}{240}\dfrac{1000 \text{ mm}}{1 \text{ m}} = 25.4 \text{ mm}$

Live load deflection (short term) should be limited to:

$$\delta_{aST} = \frac{l}{360}$$

$= \dfrac{20 \text{ ft}}{360}\dfrac{12 \text{ in}}{1 \text{ ft}} = 0.667 \text{ in}$ $\qquad\qquad$ $= \dfrac{6.10 \text{ m}}{360}\dfrac{1000 \text{ mm}}{1 \text{ m}} = 16.9 \text{ mm}$

Note that we are using subscript *a* to denote 'allowable'.

Because $\delta_{TOT} \le \delta_{aTOT}$, and $\delta_{ST} \le \delta_{aL}$, our beam is acceptable for deflection.

Step 7: Summary

In summary, our structural framing layout is as shown in Figure 4.14.

We are using Southern Pine as follows:

Sawn lumber	Glued laminated timber
8 × 24 in (200 × 620 mm) beam	6.75 × 18 in (170 × 460 mm) beam
No. 1 Grade	24F-V3 layup

4.7.2 LVL Example

Let's now design a floor joist for the framing layout shown in Figure 4.14.

Step 1: Draw Structural Layout

Key dimensional data are:

l = 16 ft l = 4.88 m
l_t = 16 in l_t = 400 mm

Step 2: Determine Loads

Step 2a: Unit Loads

The unit loads are the same as the last example

Step 2b: Load Combination

The line load, w, is substantially less than the beam, because the tributary width is much less:

$$w = (q_D + q_L)l_t$$

$$= \left(20\frac{lb}{ft^2} + 50\frac{lb}{ft^2}\right)\frac{16\ in}{12\ in/ft} \qquad = \left(0.958\frac{kN}{m^2} + 2.394\frac{kN}{m^2}\right)0.4\ m$$

$$= 93.3\frac{lb}{ft} \qquad\qquad\qquad\qquad = 1.34\frac{kN}{m}$$

Step 2c: Determine Member Moment

We are only concerned with the maximum moment, which occurs at the middle. Using the formulas in Appendix 6, we see:

$$M = \frac{wl^2}{8}$$

$$= \frac{93.3\ lb/ft\ (16\ ft)^2}{8}\frac{1\ k}{1000\ lb} \qquad = \frac{1.34\ kN/m\ (4.88\ m)^2}{8}$$

$$= 2.99\ k-ft \qquad\qquad\qquad = 3.99\ kN-m$$

It goes faster the second time!

Step 3: Material Parameters

We will design this joist using LVL material. Using Table A2.6 and choosing 1.9E WS, we get a bending reference design stress of:

$$F_b = 2,600 \text{ lb/in}^2 \qquad\qquad F_b = 17,926 \text{ kN/m}^2$$

We now apply the adjustment factors to these, following the list in Table 2.6. We will use the following factors:

Factor	Description	Source
$C_D = 1.0$	Load duration—floor load	Table A4.1
$C_M = 1.0$	Wet service—dry, interior condition	Table A4.2
$C_t = 1.0$	Temperature—sustained temperatures don't exceed 100°F (37.8°C)	Table A4.4
$C_L = 1.0$	Beam stability—beam is fully braced	Eqn (4.1)
$C_V = 1.0$	Volume—assuming 12 in (305 mm) deep	Table A4.7
$C_r = 1.15$	Repetitive member—frequently spaced joists	Section 2.4.11

Multiplying these together with the reference design values, we get the adjusted design values:

$$F'_b = F_b C_D C_M C_t C_L C_V C_r$$

$$= 2,600 \frac{\text{lb}}{\text{in}^2}(1.15) = 2,990 \frac{\text{lb}}{\text{in}^2} \qquad = 17,926 \frac{\text{kN}}{\text{m}^2}(1.15) = 20,615 \frac{\text{kN}}{\text{m}^2}$$

$$E = 1,900,000 \frac{\text{lb}}{\text{in}^2} \qquad\qquad E = 13,100 \frac{\text{MN}}{\text{m}^2}$$

$$E' = E C_M C_t$$

$$= 1,900,000 \frac{\text{lb}}{\text{in}^2}(1.0) \qquad\qquad E' = 13,100 \frac{\text{MN}}{\text{m}^2}(1.0)$$

$$= 1,900,000 \frac{\text{lb}}{\text{in}^2} \qquad\qquad = 13,100 \frac{\text{MN}}{\text{m}^2}$$

Step 4: Initial Size

Following the Initial Beam Sizing box, we will say the beam depth is equal to half the span in inches (or multiply the span in meters by 40 to get depth in millimeters).

$$d_{est} = \frac{l}{2}$$

$$= \frac{16}{2} = 8 \text{ in}$$

$$d_{est} = 40 \text{ l}$$

$$= 40(4.88) = 195 \text{ mm}$$

Using the section properties table in Appendix 1 (Table A1.3), let's try a 7.25 in (185 mm) deep member, giving us:

$d = 7.25$ in

$b = 1.75$ in

$d = 184$ mm

$b = 44$ mm

Step 5: Stress

Now that we have our size, we can calculate the section modulus, or look it up from Table A1.1.

$$S = \frac{1}{6}bd^2$$

$$S = \frac{1}{6}(1.75 \text{ in})(7.25)^2$$

$$= 15.3 \text{ in}^3$$

$$S = \frac{1}{6}(44 \text{ mm})(184 \text{ mm})^2$$

$$= 2.48 \times 10^5 \text{ mm}^3$$

We then calculate stress, adjusting the units:

$$f_b = \frac{M}{S}$$

$$= \frac{2.99 \text{ k} - \text{ft}}{15.3 \text{ in}^3}\left(12\frac{\text{in}}{\text{ft}}\right)1000\frac{\text{lb}}{\text{k}}$$

$$= 2,345 \frac{\text{lb}}{\text{in}^2}$$

$$= \frac{3.99 \text{ kN} - \text{m}}{0.248 \times 10^6 \text{ mm}^3\left(1 \text{ m}/1000 \text{ mm}\right)^3}$$

$$= 16,088 \frac{\text{kN}}{\text{m}^2}$$

This is less than the allowable stress, and so we know we are OK for stress.

Step 6: Deflection

Now, let's check deflection (serviceability). Recall the general equation for deflection is:

$$\delta_{TOT} = K_{CD}\delta_{LT} + \delta_{ST}$$

$$K_{cr} = 1.5$$

$$\delta_{LT} = \frac{5w_D l^4}{384E'I}$$

$$\delta_{ST} = \frac{5w_L l^4}{384E'I}$$

Paul W. McMullin

Pausing again, we calculate the necessary terms.

$$w_D = q_D l_t$$

$$= 20 \frac{\text{lb}}{\text{ft}^2} \left(\frac{16 \text{ in}}{12 \text{ in/ft}} \right) = 26.7 \frac{\text{lb}}{\text{ft}} \qquad = 0.958 \frac{\text{kN}}{\text{m}^2} 0.4 \text{ m} = 0.383 \frac{\text{kN}}{\text{m}}$$

$$w_L = q_L l_t$$

$$= 50 \frac{\text{lb}}{\text{ft}^2} \left(\frac{16 \text{ in}}{12 \text{ in/ft}} \right) = 66.7 \frac{\text{lb}}{\text{ft}} \qquad = 2.39 \frac{\text{kN}}{\text{m}^2} 0.4 \text{ m} = 0.956 \frac{\text{kN}}{\text{m}}$$

We'll calculate the moment of inertia this time:

$$I = \frac{1}{12} b d^3$$

$$I = \frac{1}{12} (1.75 \text{ in})(7.25 \text{ in})^3 \qquad I = \frac{1}{12} (44 \text{ mm})(184 \text{ mm})^3$$

$$= 55.6 \text{ in}^4 \qquad\qquad = 22.84 \times 10^6 \text{ mm}^4$$

With these, we can now calculate deflection:

$$\delta_{LT} = \frac{5(26.7 \text{ lb/ft})(16 \text{ ft})^4}{384(1{,}900{,}000 \text{ lb/in}^2)55.6 \text{ in}^4} \left(12 \frac{\text{in}}{\text{ft}} \right)^3$$

$$= 0.373 \text{ in}$$

$$\delta_{LT} = \frac{5(0.383 \text{ kN/m})(4.88 \text{ m})^4}{384(13{,}100 \text{ MN/m}^2)22.84 \times 10^6 \text{ mm}^4} \frac{\left(1{,}000 \text{ mm} \big/ 1 \text{ m} \right)^5}{\left(1{,}000 \text{ kN} \big/ 1 \text{ MN} \right)}$$

$$= 9.45 \text{ mm}$$

$$\delta_{ST} = \frac{5(66.7 \text{ lb/ft})(16 \text{ ft})^4}{384(1{,}900{,}000 \text{ lb/in}^2)55.6 \text{ in}^4} \left(12 \frac{\text{in}}{\text{ft}} \right)^3$$

$$= 0.931 \text{ in}$$

$$\delta_{ST} = \frac{5(0.956 \text{ kN/m})(4.88 \text{ m})^4}{384(13{,}100 \text{ MN/m}^2)22.84 \times 10^6 \text{ mm}^4} \frac{\left(1{,}000 \text{ mm} \big/ 1 \text{ m} \right)^5}{\left(1{,}000 \text{ kN} \big/ 1 \text{ MN} \right)}$$

$$= 23.6 \text{ mm}$$

$$\delta_{TOT} = 1.5(0.373 \text{ in}) + 0.931 \text{ in} \qquad \delta_{TOT} = 1.5(9.45 \text{ mm}) + 23.6 \text{ mm}$$
$$= 1.49 \text{ in} \qquad\qquad\qquad = 37.8 \text{ mm}$$

Calculating the allowable deflections, we get:

$$\delta_{aTOT} = \frac{l}{240} = \frac{16 \text{ ft}}{240}\left(\frac{12 \text{ in}}{1 \text{ ft}}\right) = 0.80 \text{ in} \qquad = \frac{4.88 \text{ m}}{240}\frac{1000 \text{ mm}}{1 \text{ m}} = 20.3 \text{ mm}$$

And live load deflection to:

$$\delta_{aL} = \frac{l}{360} = \frac{16 \text{ ft}}{360}\left(\frac{12 \text{ in}}{1 \text{ ft}}\right) = 0.53 \text{ in} \qquad \delta_{aL} = \frac{4.88 \text{ m}}{360}\frac{1000 \text{ mm}}{1 \text{ m}} = 13.6 \text{ mm}$$

Comparing these with the allowables above, we see our joist is deflecting too much. Let's select the next joist depth:

$d = 9.5$ in $\qquad\qquad\qquad d = 241$ mm

$b = 1.75$ in $\qquad\qquad\qquad b = 44$ mm

$$I = \frac{1}{12}bd^3$$

$$= \frac{1}{12}(1.75 \text{ in})(9.5 \text{ in})^3 = 125 \text{ in}^4 \qquad = \frac{1}{12}(44 \text{ mm})(241 \text{ mm})^3$$
$$= 51.3 \times 10^6 \text{ mm}^4$$

Recalculating deflection using the equations above, we get:

$\delta_{LT} = 0.166$ in $\qquad\qquad\qquad \delta_{LT} = 4.21$ mm

$\delta_{ST} = 0.414$ in $\qquad\qquad\qquad \delta_{ST} = 10.52$ mm

$\delta_{TOT} = 0.662$ in $\qquad\qquad\qquad \delta_{TOT} = 16.83$ mm

These are below the maximum allowed.

Step 7: Summary

In summary, our structural framing layout is shown in Figure 4.14. We are using 1.9E WS LVL material, 1¾ × 9½ in (44 × 241 mm) size.

Step 8: Additional Thoughts

What would happen if we forgot to brace the joists? Or perhaps someone places a bundle of plywood on the joists before they have bridging or sheathing installed. We capture this effect by calculating the beam stability factor using the following equation:

$$C_L = \frac{1 + \left(\frac{F_{be}}{F_b^*}\right)}{1.9} - \sqrt{\left[\frac{1 + \left(\frac{F_{be}}{F_b^*}\right)}{1.9}\right]^2 - \frac{\frac{F_{be}}{F_b^*}}{0.95}}$$

It's an intimidating equation, but we'll walk our way through it.

$$F_{bE} = \frac{120E'_{min}}{R_B^2} \quad \text{, related to lateral torsional buckling strength}$$

$$R_B = \sqrt{\left(\frac{l_e d}{b^2}\right)} \quad \text{, a slenderness parameter}$$

We will conservatively take l_e as twice the unbraced length, yielding:

$$l_e = 2l$$

$$= 2(16 \text{ ft})\frac{12 \text{ in}}{1 \text{ ft}} = 384 \text{ in} \qquad = 2(4.88 \text{ m})\frac{1000 \text{ mm}}{1 \text{ m}} - 9{,}760 \text{ mm}$$

$$R_B = \sqrt{\frac{384 \text{ in} (9.5 \text{ in})}{(1.75 \text{ in})^2}} = 34.5 \qquad R_B = \sqrt{\frac{9{,}760 \text{ mm} (241 \text{ mm})}{(44 \text{ mm})^2}} = 34.9$$

This is less than 50, and so we are good to continue. If it was greater than 50, the resulting C_L would be so low that the beam would have very little capacity.

Finding E_{min}:

$$E_{min} = 966{,}000 \frac{\text{lb}}{\text{in}^2} \qquad E_{min} = 6{,}660 \frac{\text{MN}}{\text{m}^2}$$

$$E'_{min} = E_{min}C_M C_t$$

$$= 966{,}000 \frac{\text{lb}}{\text{in}^2}(1.0) = 966{,}000 \frac{\text{lb}}{\text{in}^2} \qquad = 6{,}660 \frac{\text{MN}}{\text{m}^2}(1.0) = 6{,}660 \frac{\text{MN}}{\text{m}^2}$$

We can now calculate F_{bE}:

$$F_{bE} = \frac{1.20\left(966{,}000 \text{ lb/in}^2\right)}{(34.5)^2} \qquad F_{bE} = \frac{1.20\left(6{,}660 \text{ MN/m}^2\right)}{(34.9)^2}\frac{1000 \text{ kN}}{1 \text{ MN}}$$

$$= 973.9 \frac{\text{lb}}{\text{in}^2} \qquad\qquad = 6{,}560 \frac{\text{kN}}{\text{m}^2}$$

This is less than half the allowable bending stress, assuming a fully braced joist.

F^*_b is F_b times all the applicable adjustment factors except C_L:

$$F^*_b = F_b C_D C_M C_t C_V C_r$$

$$= 2,600 \frac{\text{lb}}{\text{in}^2}(1.0)1.15 = 2,990 \frac{\text{lb}}{\text{in}^2} \qquad = 17,926 \frac{\text{kN}}{\text{m}^2}(1.0)1.15 = 20,615 \frac{\text{kN}}{\text{m}^2}$$

This gives us the pieces of the puzzle to calculate C_L. It helps to calculate the individual portions of the equation.

$$C_L = \frac{1+\left(973.9 \big/ 2,290\right)}{1.9} - \sqrt{\left(\frac{1+\left(973.9 \big/ 2,290\right)}{1.9}\right)^2 - \frac{973.9 \big/ 2,290}{0.95}}$$

$$= 0.32$$

$$C_L = \frac{1+\left(6,560 \big/ 20,615\right)}{1.9} - \sqrt{\left(\frac{1+\left(6,560 \big/ 20,615\right)}{1.9}\right)^2 - \frac{6,560 \big/ 20,615}{0.95}}$$

$$= 0.31$$

And, calculating the new allowable bending stress:

$$F'_b = F^*_b C_L$$

$$= 2,990 \frac{\text{lb}}{\text{in}^2}(0.32) = 957 \frac{\text{lb}}{\text{in}^2} \qquad = 20,615 \frac{\text{kN}}{\text{m}^2}(0.31) = 6,391 \frac{\text{kN}}{\text{m}^2}$$

This is quite a reduction in strength and indicates that the joist fails.

Now, to help you relax and fall asleep, see what size of LVL it would take to carry the same loads.

4.7.3 I-Joist Example

Continuing our floor design, let's look at what it takes to make the the floor out of I-joists. Following the layout and loads from the LVL example, we continue on step 3.

Step 3: Material Parameters

I-joists are different than other materials. We look at bending capacity (i.e., moment) rather than allowable stresses. For this, we focus on M', rather than F'_b.

Paul W. McMullin

The applicable adjustment factors from Table 2.7 are shown in the table.

Factor	Description	Source
$C_D = 1.0$	Load duration—floor load	Table A4.1
$C_M = 1.0$	Wet service—dry, interior condition	Table A4.2
$C_t = 1.0$	Temperature—sustained temperatures don't exceed 100°F (37.8°C)	Table A4.4
$C_L = 1.0$	Beam stability—I-joists must be fully braced	Eqn. (4.1)
$C_r = 1.15$	Repetitive member—frequently spaced joists	Section 2.4.11

Step 4: Initial Size

We skip this step and go directly to choosing the joist we need, based on the moment.

Step 5: Strength

Going to Table A2.7, we find a joist size that has a higher allowable moment than our demand. We see a 5000–1.8 series joist, 11⅞ in (302 mm) deep, has sufficient strength. Assuming a 12 in (305 mm) structural depth is acceptable (which it commonly is), we proceed with our calculations. If we need to restrict the joist depth to 9½ in (240 mm), we could use the 6000–1.8 series joists.

$M_r = 3{,}150$ lb $M_r = 4{,}721$ N-m

Multiplying this by the adjustment factors, we get the adjusted design strength:

$$M'_r = M_r C_D C_M C_t C_L C_r$$

= 3,150 lb-ft (1.0)1.15   = 4,721 N-m (1.0)1.15
= 3,623 lb-ft   = 5,429 N-m

As this is larger than the required moment, we are OK.

Step 6: Deflection

Moving on to deflection, joist manufacturers give us a combined material and geometric stiffness, EI.

$EI = 250 \times 10^2$ in²lb $EI = 717 \times 10^2$ mm²kN

Unusual units, but the result will make more sense.

Adjusting to get *EI'*, we have:

$EI' = EIC_M C_t$
= 250 × 10⁶ in²lb (1.0)   = 717 × 10⁶ mm²kN (1.0)
= 250 × 10⁶ in²lb   = 717 × 10⁶ mm²kN

$EI' = EIC_M C_t$
$= 250 \times 10^6 \text{ in}^2\text{lb } (1.0) \qquad = 717 \times 10^6 \text{ mm}^2\text{kN } (1.0)$
$= 250 \times 10^6 \text{ in}^2\text{lb} \qquad = 717 \times 10^6 \text{ mm}^2\text{kN}$

Following the equations and line loads of the previous examples, and *EI'* from above, we get:

$\delta_{LT} = 0.157 \text{ in} \qquad\qquad \delta_{LT} = 3.95 \text{ mm}$

$\delta_{ST} = 0.393 \text{ in} \qquad\qquad \delta_{ST} = 9.86 \text{ mm}$

$\delta_{TOT} = 0.629 \text{ in} \qquad\qquad \delta_{TOT} = 15.78 \text{ mm}$

These are lower than the allowables. Our joist works!

Step 7: Summary

We are using a 11⅞ in (300 mm) joist, 5000–1.8 series.

4.8 WHERE WE GO FROM HERE

This chapter thoroughly presents the criteria for simply supported beams. For multi-span beams, we need to investigate tension in the top over the supports. For glued laminated timber, this would require a different **lam** layup.

Moving forward, we look into the shear behavior of beams—outlined in Chapter 5.

NOTES

1. ANSI/AWC. *National Design Specification (NDS) Supplement: Design Values for Wood Construction* (Leesburg, VA: AWC, 2015).
2. ANSI/AWC. *National Design Specification (NDS) for Wood Construction* (Leesburg, VA: AWC, 2015).
3. Paul W. McMullin and Jonathan S. Price. *Introduction to Structures* (New York: Routledge, 2016).

Paul W. McMullin

Timber Shear

Chapter 5

Paul W. McMullin

Shear is fundamental to how beams resist load. It holds the outer layers of a bending member together, causing them to act as one. This chapter will focus on how we size beams for shear stress. Chapter 8 covers shear wall design to resist wind and seismic forces.

Let's take a moment and conceptually understand the fundamentals of shear behavior. You are likely familiar with the action scissors make when cutting paper or fabric. The blades are perpendicular to the material, going in opposite directions. This creates a tearing of the material like that shown in Figure 5.1. In beam shear, the action is similar, but the movement of material is parallel to the length of the beam. The top portion moves relative to the bottom, illustrated in Figure 5.2.

Shear strength is fundamentally tied to bending strength and stiffness. If we take a stack of paper and lay it across two supports, it sags (Figure 5.3a), unable to carry even its own load. If we glue each strip of paper together, we get a beam with enough strength and stiffness to carry a reasonable load, as shown in Figures 5.3b and 5.3c. And so it is with wood beams. The lengthwise fibers provide bending strength, but it is the lignin between the fibers that is the glue, causing the fibers to work together.

5.1 STABILITY

Shear stability is not a concern in solid timber members. The sections are compact enough to not experience shear buckling.

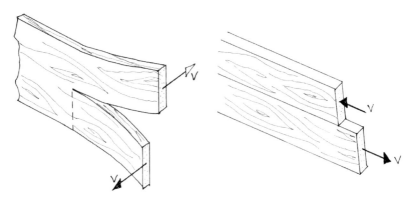

Figure 5.1 Shearing action similar to scissors

Figure 5.2 Shearing action from bending

Paul W. McMullin

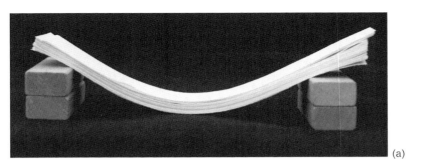

Figure 5.3 Paper beam with layers (a) unglued and (b, c) glued (Lego™ figures courtesy Peter McMullin)

For I-joists, the web may buckle under point loads or end **reactions**. Though this is really a column-buckling-type behavior, it applies to the design of bending members. In these cases, it is necessary to place blocking to stiffen the web of the joists, as shown in Figure 5.4. Note the gap at the top of the block to accommodate thermal and moisture movement.

5.2 CAPACITY

Finding the shear capacity of the wood is a two-step process. First, we look up the reference design stress, F_v, and then we adjust this with the applicable adjustment factors (e.g., C_D, C_t, C_m), yielding F'_v.

5.2.1 Reference Design Values

Shear reference design values are a function of species, but usually independent of grade. For example, the reference design shear stress of Redwood is 160 lb/in² (1,103 kN/m²) for any grades from Stud to Clear Structural, even though the bending values for these are drastically different. This is because shear strength comes from lignin, which glues the lengthwise fibers (cellulose) together.

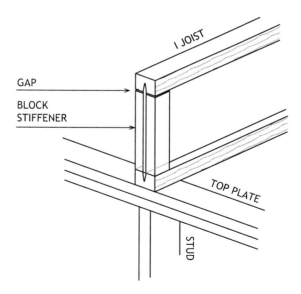

Figure 5.4 I-joist stiffening at end support

Paul W. McMullin

As shear strength rarely changes within a species, it is likely to control beam design of higher grade woods, rather than lower.

To find the reference design shear stress, we begin by first deciding what wood product we will use—sawn lumber, glued laminated timber, SCL, or I-joists. This will lead us to one of the following tables in Appendix 2, extracted from the *NDS Supplement*.[1]

- visually graded dimension lumber—Table A2.1;
- visually graded timbers (5 × 5 in and larger)—Table A2.2;
- mechanically graded Douglas Fir–Larch (North)—Table A2.3;
- visually graded Southern Pine—Table A2.4;
- glued laminated timber—Table A2.5;
- SCL—Table A2.6;
- I-joists—Table A2.7.

We find the reference design value as follows:

- In the first column, select the species. It is best if this is locally available and sustainably harvested.
- Under the species heading, select the grade. No. 2 is very common in Douglas Fir and Southern Pine. Consider calling lumber suppliers in the vicinity of the project to see what they commonly stock.
- In the second column, choose the Size Class, if available.
- Finally, find the column that corresponds to the property you are looking for. Read down to the row for the species and grade you are using and read the value in lb/in^2 (kN/m^2).

For example, say you want to know the reference design shear stress for Southern Pine. You would enter Table A2.4 and go across the column heading until you see F_v. Scanning down the column, you would see it is the same for all grades—175 lb/in^2 (1,207 kN/m^2).

5.2.2 Adjusted Design Values

With the reference design shear stress, F_v, we now adjust it for the factors discussed in Chapter 2. Following Tables 2.4–2.7, we can quickly see which factors apply to shear for different wood products. In equation form:

$$F'_v = F_v C_D C_M C_t C_i \text{ for sawn lumber} \tag{5.1}$$

$$F'_v = F_v C_D C_M C_t C_{vr} \text{ for glued laminated timber} \tag{5.2}$$

$$F'_v = F_v C_D C_M C_t \text{ for SCL} \tag{5.3}$$

$$V'_t = V_t C_D C_M C_t \text{ for I-Joists} \tag{5.4}$$

where:

F_v = reference design shear strength from Appendix 2

C_x = adjustment factors; see Table 2.3

Notice that for I-joists we are working in capacity units of force. This is because the manufacturers give shear strength, instead of the necessary

Table 5.1 **Shear strength for varying species and wood types**

Imperial Units	Aspen	Spruce–Pine–Fir	Redwood	Southern Pine	Douglas Fir–Larch (N)	Glue–Lam	LVL	Glue–Lam
Shear Strength V (k)								
F_v (lb/in²)								
	120	135	160	175	180	265	285	300
2 × 4	0.42	0.47	0.56	0.61	0.63	0.93	1.00	1.05
2 × 6	0.66	0.74	0.88	0.96	0.99	1.46	1.57	1.65
2 × 8	0.87	0.98	1.16	1.27	1.31	1.92	2.07	2.18
2 × 10	1.11	1.25	1.48	1.62	1.67	2.45	2.64	2.78
2 × 12	1.35	1.52	1.80	1.97	2.03	2.98	3.21	3.38
6 × 12	5.06	5.69	6.75	7.38	7.59	11.2	12.0	12.7
6 × 16	6.82	7.67	9.09	9.95	10.2	15.1	16.2	17.1
6 × 20	8.58	9.65	11.4	12.5	12.9	18.9	20.4	21.5
6 × 24	10.3	11.6	13.8	15.1	15.5	22.8	24.6	25.9
10 × 16	11.8	13.3	15.7	17.2	17.7	26.0	28.0	29.5
10 × 20	14.8	16.7	19.8	21.6	22.2	32.7	35.2	37.1
10 × 24	17.9	20.1	23.8	26.0	26.8	39.4	42.4	44.7
14 × 18	18.9	21.3	25.2	27.6	28.4	41.7	44.9	47.3
14 × 24	25.4	28.6	33.8	37.0	38.1	56.0	60.3	63.5

Note: Apply appropriate adjustment factors

Paul W. McMullin

variables to calculate shear stress. We therefore compare shear force, V, with adjusted design shear strength, V'_r. Additionally, we are not using the LRFD adjustment factors, as we are working in allowable stress units.

To assist you in preliminary sizing for shear, Table 5.1 presents shear capacities for various section sizes and materials.

Metric Units	Shear Strength V (kN)							
	Aspen	Spruce–Pine–Fir	Redwood	Southern Pine	Douglas Fir–Larch (N)	Glue-Lam	LVL	Glue-Lam
				F_v (kN/m²)				
	827	931	1,103	1,207	1,241	1,827	1,965	2,068
2 × 4	1.87	2.10	2.49	2.72	2.80	4.13	4.44	4.67
2 × 6	2.94	3.30	3.91	4.28	4.40	6.48	6.97	7.34
2 × 8	3.87	4.35	5.16	5.64	5.80	8.55	9.19	9.67
2 × 10	4.94	5.55	6.58	7.20	7.41	10.9	11.7	12.3
2 × 12	6.01	6.76	8.01	8.76	9.01	13.3	14.3	15.0
6 × 12	22.5	25.3	30.0	32.8	33.8	49.7	53.5	56.3
6 × 16	30.3	34.1	40.4	44.2	45.5	67.0	72.1	75.8
6 × 20	38.2	42.9	50.9	55.7	57.2	84.3	90.6	95.4
6 × 24	46.0	51.7	61.3	67.1	69.0	102	109	115
10 × 16	52.4	59.0	69.9	76.4	78.6	116	124	131
10 × 20	65.9	74.2	87.9	96.1	98.9	146	157	165
10 × 24	79.4	89.4	106	116	119	175	189	199
14 × 18	84.1	94.6	112	123	126	186	200	210
14 × 24	113	127	151	165	169	249	268	282

Note: Apply appropriate adjustment factors

5.3 DEMAND VS. CAPACITY

Once we have the adjusted design stress, F'_v, we compare it with the actual shear stress. For a simply supported beam, shear is at its maximum at the end of the beam. Inside the member, stress varies from zero at the top and bottom surface to a maximum at the middle, illustrated in Figure 5.5. Exploring this further, we see how shear stress changes along the length of several beams in Figure 5.6. In the single-span, simply supported beam, the shear stress is zero at the middle and maximum at the ends. A cantilever is the opposite, with maximum stress at the supported end. A multi-span beam has maximum shear stress at the supports.

Regardless of the shear force variation along the beam length, we find shear stress from equation (5.5) for any section type. For rectangular sections, we can use equation (5.6). Remember to watch your units.

$$f_v = \frac{VQ}{Ib} \quad \text{for any section} \tag{5.5}$$

$$f_v = \frac{3V}{2bd} \quad \text{for rectangular sections} \tag{5.6}$$

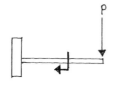

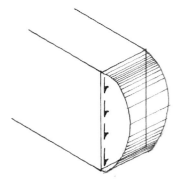

Figure 5.5 Shear stress distribution in cross section

Paul W. McMullin

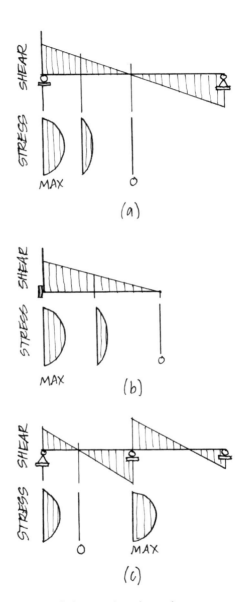

Figure 5.6 Shear stress variation at points along a beam

where:

V = shear force at location of interest, k (kN)

Q = first moment about the neutral axis, in^3 (mm^3)

I = moment of inertia, in^4 (mm^4)

b = section width, in (mm)

d = section depth, in (mm)

Because timber beams usually have a constant cross section (prismatic) along their length, we are typically concerned with the maximum shear force (and therefore stress). However, using the equations above, we can insert the shear force, V, at any point along the beam and find its corresponding stress. Similar to concrete beams, for uniformly **distributed** loads, we may design for the shear at a distance d away from the support—giving us a slight advantage.

If the shear stress, f_v, is less than the adjusted design stress, F'_v, we are OK. If it is higher, we select a deeper or wider beam and recalculate.

To help you select a beam large enough to carry a given shear load, Table 5.1 provides the shear strength of various beam sizes for a number of wood species. For I-joists, see Table A2.7. Remember to adjust these tables with the necessary factors, C_D, C_M, and C_t.

5.4 DEFLECTION

Shear action contributes little to bending deflection—usually only 3–5 percent. The modulus of elasticity values, E, for sawn lumber and glued laminated timber are slightly reduced to account for shear deflection. SCL (LVL and I-joists) do not have an adjustment on modulus of elasticity (though it is coming). If shear deflection is a concern—which it would be for short beams with high point loads—we can either just increase the deflection 3–5 percent or use an equation that takes shear deflection into account. ICC reports for SCL typically provide these equations for uniform and point loads.

For a simply supported beam, with a uniform load, this equation is:

$$\delta = \frac{5wl^4}{384EI} + \frac{0.96wl^2}{0.4EA}$$

(5.7)

Bending + Shear

Paul W. McMullin

where:

w = uniform distributed load, lb/ft (kN/m)

l = beam span ft (m)

E = modulus of elasticity, lb/in^2 (kN/m^2)

I = moment of inertia, in^4 (mm^4)

A = cross-sectional area, in^2 (mm^2)

Refer to Chapter 4 for additional discussion on deflection calculations and acceptance criteria.

5.5 DETAILING

Detailing considerations for shear are similar to those found in Chapter 9 for connections, and in Chapter 4 for bending. Keep in mind that shear stress is at a maximum near supports and at the mid-height of the cross section. Any field modifications for piping or conduit in these areas will reduce shear strength and should be carefully evaluated by a professional engineer.

5.6 DESIGN STEPS

1. Draw the structural layout; include span dimensions and tributary width.
2. Determine loads:
   (a) unit loads;
   (b) load combinations yielding a line load;
   (c) member end shear.
3. Material parameters—find reference design values and adjustment factors.
4. Estimate initial size or use size from bending and deflection check.
5. Calculate stress and compare with adjusted design stress.
6. Summarize the results.

5.7 DESIGN EXAMPLE

5.7.1 Beam Shear Example

Building on the beam example in Chapter 4, we will now check the shear capacity of the sawn beam shown in Figure 5.7. We leave it to you to check the glued laminated beam.

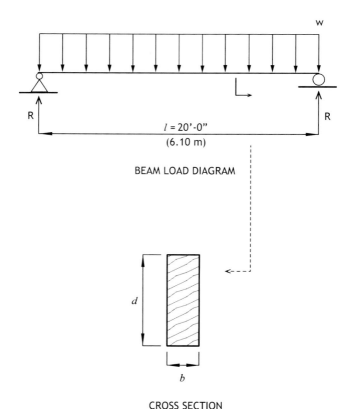

BEAM LOAD DIAGRAM

CROSS SECTION

Figure 5.7 Example beam free body diagram and cross section

Step 1: Draw Structural Layout

Begin by drawing the framing layout, complete with the dimensions needed in the design, shown in Figure 5.8. Key dimensional data are length, l, and tributary width, l_t.

l = 20 ft	l = 6.10 m
l_t = 16 ft	l_t = 4.88 m

Paul W. McMullin

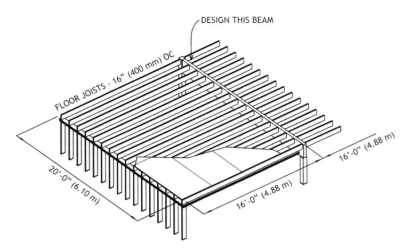

Figure 5.8 Example framing configuration

Step 2: Determine Loads

Step 2a: Unit Loads

The unit dead and live load are:

$$q_D = 20\,\frac{\text{lb}}{\text{ft}^2} \qquad\qquad q_D = 0.958\,\frac{\text{kN}}{\text{m}^2}$$

$$q_L = 50\,\frac{\text{lb}}{\text{ft}^2} \qquad\qquad q_L = 2.394\,\frac{\text{kN}}{\text{m}^2}$$

Step 2b: Load Combination

Because this is a floor, the live load dominant combination will control. Multiplying it by the tributary width, we find the line load, w, as:

$$w = (q_D + q_L)l_t$$

$$= \left(20\,\frac{\text{lb}}{\text{ft}^2} + 50\,\frac{\text{lb}}{\text{ft}^2}\right)16\text{ ft} \qquad\qquad = \left(0.958\,\frac{\text{kN}}{\text{m}^2} + 2.394\,\frac{\text{kN}}{\text{m}^2}\right)4.88\text{ m}$$

$$= 1{,}120\,\frac{\text{lb}}{\text{ft}} \qquad\qquad\qquad\qquad\qquad = 16.4\,\frac{\text{kN}}{\text{m}}$$

We are only concerned with the maximum shear, which occurs at the ends. Using the formulas in Appendix 6 and Figure 4.15, we see

$$V = \frac{wl}{2}$$

$$= \frac{1{,}120 \text{ lb/ft } (20 \text{ ft})}{2} = 11.2 \text{ k} \qquad = \frac{16.4 \text{ kN/m } (6.10 \text{ m})}{2} = 50.0 \text{ kN}$$

Half the total beam load is supported at each end.

Step 3: Material Parameters

Looking at the sawn lumber beam and using Table A2.4, we will find the reference design shear stress for Southern Pine, No. 1.

$$F_v = 175 \frac{\text{lb}}{\text{in}^2} \qquad\qquad F_v = 1{,}207 \frac{\text{kN}}{\text{m}^2}$$

We now apply the adjustment factors to these, following the list in Table 2.4. We will use the factors shown in the table below.

Factor	Description	Source
$C_D = 1.0$	Load duration—floor load	Table A4.1
$C_M = 1.0$	Wet service—dry, interior condition	Table A4.2
$C_t = 1.0$	Temperature—sustained temperatures don't exceed 100°F (37.8°C)	Table A4.4
$C_i = 1.0$	Incising—not treated wood	Table A4.10

We multiply these together with the reference design stress to obtain the adjusted design stress:

$$F'_v = F_v C_D C_M C_t C_i$$

$$= 175 \frac{\text{lb}}{\text{in}^2} (1.0) = 175 \frac{\text{lb}}{\text{in}^2} \qquad = 1{,}207 \frac{\text{kN}}{\text{m}^2} (1.0) = 1{,}207 \frac{\text{kN}}{\text{m}^2}$$

Step 4: Initial Size

We already have sizes from the bending example. Using those:

$$d = 23.5 \text{ in} \qquad\qquad d = 600 \text{ mm}$$
$$b = 7.5 \text{ in} \qquad\qquad b = 190 \text{ mm}$$

Paul W. McMullin

Step 5: Stress

With that assumed beam size, we calculate horizontal shear stress using the following equation:

$$f_v = \frac{3V}{2bd}$$

$$= \frac{3(11.2\ \text{k})}{2(7.5\ \text{in})23.5\ \text{in}}\frac{1000\ \text{lb}}{1\ \text{k}}$$

$$= 95.3\frac{\text{lb}}{\text{in}^2}$$

$$= \frac{3(50.0\ \text{kN})}{2(190\ \text{mm})600\ \text{mm}}\left(\frac{1000\ \text{mm}}{1\ \text{m}}\right)^2$$

$$= 658\frac{\text{kN}}{\text{m}^2}$$

Comparing this with the adjusted design shear stress, we see our beam is OK. That's it!

Step 6: Summary

Our sawn beam in Chapter 4 works.

5.7.2 LVL Joist Shear Example

Let's now check the LVL floor joist from Chapter 4 for shear. Picking up at step 2c, we calculate the shear force in the joist.

Step 2c: Determine Member Shear

Find the maximum shear (at the ends):

$$V = \frac{wl}{2}$$

$$= \frac{93.3\ \text{lb/ft}(16\ \text{ft})}{2}\frac{1\ \text{k}}{1000\ \text{lb}}$$

$$= 0.75\ \text{k}$$

$$= \frac{1.34\ \text{kN/m}(4.88\ \text{m})}{2}$$

$$= 3.27\ \text{kN}$$

Half the total beam load is supported at each end.

Step 3: Material Parameters

We will design this beam using LVL material. Using Table A2.6 and choosing 1.9E WS, we find the bending reference design stress is:

$$F_v = 285\frac{\text{lb}}{\text{in}^2}$$

$$F_v = 1{,}965\frac{\text{kN}}{\text{m}^2}$$

We now apply the adjustment factors to these, following the list in Table 2.6. We will use the following factors:

Factor	Description	Source
$C_D = 1.0$	Load duration—floor load	Table A4.1
$C_M = 1.0$	Wet service—dry, interior condition	Table A4.2
$C_t = 1.0$	Temperature—sustained temperatures don't exceed 100°F (37.8°C)	Table A4.4

Multiplying these together with the reference design stress, we get the adjusted design stress (i.e., allowable stress):

$$F'_v = F_v C_D C_M C_t$$

$$= 285 \frac{lb}{in^2}(1.0) = 285 \frac{lb}{in^2} \qquad = 1{,}965 \frac{kN}{m^2}(1.0) = 1{,}965 \frac{kN}{m^2}$$

Step 4: Initial Size

We already have sizes from the LVL bending example. Using those:

$d = 9.5$ in	$d = 241$ mm
$b = 1.75$ in	$b = 44$ mm

Step 5: Stress

With our size, we calculate horizontal shear stress using the following equation:

$$f_v = \frac{3V}{2bd}$$

$$= \frac{3(750\ lb)}{2(1.75\ in)9.5\ in} \qquad = \frac{3(3.27\ kN)}{2(44\ mm)241\ mm}\left(\frac{1000\ mm}{1\ m}\right)^2$$

$$= 68 \frac{lb}{in^2} \qquad\qquad = 463 \frac{kN}{m^2}$$

Because $f_v < F'_v$ the joist is OK for shear.

Step 6: Summary

Our LVL joist in Chapter 4 works.

5.7.3 I-Joist Shear Example

Finally, let's check the shear strength of the I-joist example in Chapter 4. Picking up at step 3, we use the data in the LVL joist bending and shear examples.

Paul W. McMullin

Step 3: Material Parameters

I-joists are different than other materials. We look at capacity rather than allowable stresses. For shear, we will be concerned with V', rather than F'_v.

The applicable adjustment factors from Table 2.7 are shown in the table.

Factor	Description	Source
$C_D = 1.0$	Load duration—floor load	Table A4.1
$C_M = 1.0$	Wet service—dry, interior condition	Table A4.2
$C_t = 1.0$	Temperature—sustained temperatures don't exceed 100°F (37.8°C)	Table A4.4

Step 4: Initial Size

We skip this step, as we already have our joist size from Chapter 4.

Step 5: Strength

Going to Table A2.7, we find the joist strength for a 5000–1.8 joist, 11⅞ in (302 mm) deep section:

$$V_r = 1.63 \text{ k} \qquad\qquad V_r = 7.228 \text{ kN}$$

Multiplying this by the adjustment factors, we get the adjusted design strength:

$$V'_r = V_r C_D C_M C_t$$
$$= 1.63 \text{ k } (1.0) = 1.63 \text{ k} \qquad = 7.23 \text{ kN } (1.0) = 7.23 \text{ kN}$$

Because $V < V'_r$, the joist is OK for shear.

Step 6: Summary

Our I-Joist in Chapter 4 works.

5.8 WHERE WE GO FROM HERE

This chapter covers shear design for the common application of all lumber types. Shear values for glued laminated timber beams that are not prismatic, subjected to impact or fatigue loads, or have notches, must be reduced by the shear reduction factor, C_{vr} (0.72).

When we design built-up beams, we pay particular attention to the horizontal shear where members interface. We need to ensure we have enough fasteners to fully connect the built-up pieces.

NOTE

1. ANSI/AWC. *National Design Specification (NDS) Supplement:* *Design Values for Wood Construction* (Leesburg, VA: AWC, 2015).

Timber Compression

Chapter 6

Paul W. McMullin

Columns make open space possible. Without them, we would be subject to the limitations of walls. Columns range from round poles to rectangular sawn sections to structural composite lumber of a variety of shapes, sizes, and layups. The discussion of columns applies equally to truss compression members.

Historically, columns were solid sawn (Figure 6.1) or of built-up sections. Today, we find solid sawn columns, along with composite lumber columns, like that in Figure 6.2.

Common timber columns range from 4 × 4 in (100 mm) to 24 × 24 in (600 mm). Column lengths range from 8 to 20 ft (2.4–6 m). Larger columns and lengths are possible—though difficult to come by.

In this chapter, we learn what parameters must be considered in the design of columns, some preliminary sizing tools, and how to do the

Figure 6.1 Timber column in historic barn, Cane River Creole National Historical Park, Natchitoches, Louisiana

Source: Photo courtesy of Robert A. Young © 2007

Figure 6.2 Structural composite lumber (PSL) column

Source: Photo courtesy of Weyerhaeuser © 2016

Paul W. McMullin

in-depth calculations. As with earlier chapters, take your time, clearly illustrate your ideas, and things will start to make sense.

6.1 STABILITY

Columns are unbraced along their length, and therefore prone to buckling. In fact, buckling is the driving consideration in column design. Because of this, the most efficient timber columns are square—providing the same buckling capacity about each axis.

To gain a conceptual understanding of column buckling, take a straw and apply a compression load at the top, as in Figure 6.3a. Note how much force it takes before the column starts bowing (buckling) out. Now, have a friend gently brace the straw in the middle so that it can't move horizontally. Apply a force until the straw starts to buckle, as shown in Figure 6.3b. Notice how much more force it takes.

Slenderness is the driving parameter for column buckling. Taller or thinner columns are more susceptible to buckling. Mathematically, we

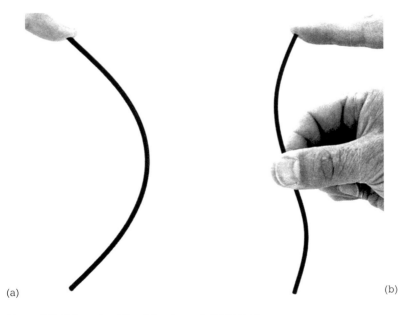

(a)　　　　　　　　　　　　　　　　　　　　　　　　　(b)

Figure 6.3 Column buckling (a) unbraced, (b) braced

represent this with length and radius of gyration. (Radius of gyration is the square root of moment of inertia divided by area.) Simplifying this somewhat, we use the following expression to measure slenderness:

$$\frac{l_e}{d}$$

(6.1)

where:

l_e = effective length, kl

d = width of smallest cross-section dimension, in (mm)

l = unbraced length, ft (m)

k = effective length factor, shown in Figure 6.4

The code limits the slenderness ratio to 50. Even if we went over this, the column stability factor, C_P, would be so small, the column would be useless.

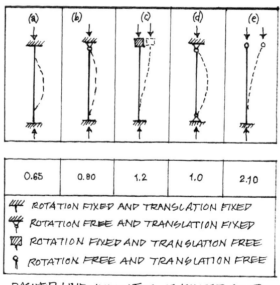

Figure 6.4 Effective length factors for simple columns

Paul W. McMullin

The column stability factor, C_P, reduces the allowable stress to account for slenderness. We calculate it utilizing the following equation:

$$C_P = \frac{1+\left(F_{cE}\big/F_c^*\right)}{2c} - \sqrt{\left[\frac{1+\left(F_{cE}\big/F_c^*\right)}{2c}\right]^2 - \frac{\left(F_{cE}\big/F_c^*\right)}{c}}$$

(6.2)

where:

F_c^* = F_c multiplied by all the adjustment factors but C_P

$$F_{cE} = \frac{0.822 E'_{min}}{\left(l_e\big/d\right)^2}$$

(6.3)

E'_{min} = adjusted minimum elastic modulus, lb/in^2 (kN/m^2)

c = 0.8 for sawn lumber, 0.85 for round poles and piles, 0.9 for glued laminated timber and SCL

Like the beam stability factor, this can be a little painful. Be patient with yourself. It helps to break up the equation.

To gain a sense of how the column stability factor varies, Figure 6.5 shows C_P for different column length and depths. Notice that, as the cross section gets smaller, the stability factor drops. Additionally, as column length gets longer, the stability factor goes down for the same depth.

6.2 CAPACITY

To find compression capacity, we obtain the reference design stress from Appendix 2 (or *NDS Supplement*)[1], and then account for numerous effects using the adjustment factors, listed in Table 2.3.

6.2.1 Reference Design Values

In compression design, we consider two actions: compression parallel and perpendicular to the grain, illustrated in Figure 6.6. We use compression parallel to the wood grain, F_c, for loads applied along the length of the column. We use compression perpendicular to grain, F'_c, values for bearing forces (such as a joist sitting on a wall). These values vary based on species and grade. Appendix 2 provides reference design values for a handful of species and grades. For column stability, we also use **minimum modulus of elasticity**, E_{min}.

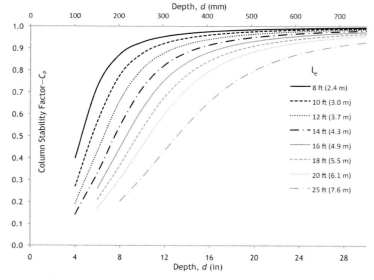

Figure 6.5 Column stability factor as a function of length and cross-section size for Southern Pine No. 2 Dense

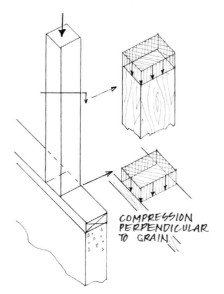

COMPRESSION PERPENDICULAR TO GRAIN

Figure 6.6 Axial and bearing stresses in a column

Paul W. McMullin

To find the reference design stress, we begin by selecting a wood product—sawn lumber, glued laminated timber, or SCL. This will lead us to one of the following tables in Appendix 2:

- Visually graded dimension lumber—Table A2.1;
- Visually graded timbers (5 × 5 in and larger)—Table A2.2;
- Mechanically graded Douglas Fir–Larch (North)—Table A2.3;
- Visually graded Southern Pine—Table A2.4;
- Glued laminated timber—Table A2.5;
- Structural Composite Lumber—Table A2.6;
- I-joists—Table A2.7.

Again, we can find the reference design stress as follows:

- In the first column, select the species.
- Under the species heading, select the grade.
- In the second column, choose the Size Class, if applicable.
- Finally, find the column that corresponds to F_c or $F_{c\perp}$. Read down to the row for the species and grade you are using and read the value in lb/in^2 (kN/m^2).

For example, assume you want to know F_c for Bald Cypress, Select Structural, at least 5 × 5 in (127 × 127 mm): you enter Table A2.2 and scan down the first column to the line that says Select Structural under Bald Cypress. You then scan to the right until you find the column with F_c in it. The number in this column is the reference design compressive stress parallel to the grain.

6.2.2 Adjusted Design Values

We now adjust the reference design values for the variables discussed in Chapter 2. Following Tables 2.4–2.6, we can see which factors apply to axial or cross-grain compression for different wood products. For axial compression, the equations are as follows:

$$F'_c = F_c C_D C_M C_t C_F C_i C_P \text{ sawn lumber} \tag{6.4}$$

$$F'_c = F_c C_D C_M C_t C_P \text{ for glued laminated timber and SCL} \tag{6.5}$$

where:

F_c = reference design compression stress from Appendix 2

C_x = adjustment factors; see Table 2.3

For cross-grain compression (bearing conditions), the equations change somewhat.

$$F'_{c\perp} = F_{c\perp}C_M C_t C_i C_b \text{ for sawn lumber} \tag{6.6}$$

$$F'_{c\perp} = F_{c\perp}C_M C_t C_b \text{ for glued laminated timber and SCL} \tag{6.7}$$

where:

$F_{c\perp}$ = reference design bearing value from Appendix 2

C_x = adjustment factors; see Table 2.3

6.3 DEMAND VS. CAPACITY

With the adjusted design stress for compression both parallel and perpendicular to the grain, we compare this with the **axial stress**. The form of the equation is the same for both, but written out separately for clarity:

$$f_c = \frac{P}{A_g} \qquad \text{compression parallel to grain} \tag{6.8}$$

$$f_{c\perp} = \frac{P}{A_{brg}} \qquad \text{compression perpendicular to grain} \tag{6.9}$$

INITIAL COLUMN SIZING

A simple rule of thumb for column size is to divide the height in feet by 2 to get the column width in inches. For metric, multiply the height in meters by 40 to get the width in millimeters. For example, a 12 ft (3.66 m) column would be a minimum of 6 in (150 mm) wide. For heavy loads, increase the size by 25–50 percent.

If we know the load to the column, we can refine our column size using Table 6.1. This provides allowable axial compression capacity in kips (kN) for varying section sizes, timber reference design stresses, and column stability factors. We compare the calculated axial load with those in the tables to get a close approximation for initial size—assuming we guess well on the column stability factor (recall Figure 6.5). Remember the effect that the adjustment factors may have—bumping size up or down accordingly.

Paul W. McMullin

Table 6.1 Axial strength of sawn lumber for varying compressive stresses

Imperial Units		Compression Strength, P (k)								
		$F_c = 700 \ lb/in^2$			$F_c = 1,400 \ lb/in^2$			$F_b = 2,500 \ lb/in^2$		
Column Size	A_g, in^2	C_P			C_P			C_P		
		0.85	0.70	0.55	0.85	0.70	0.55	0.85	0.70	0.55
5 × 5	20.25	12.0	9.9	7.8	24.1	19.8	15.6	43.0	35.4	27.8
6 × 6	30.25	18.0	14.8	11.6	36.0	29.6	23.3	64.3	52.9	41.6
8 × 8	56.25	33.5	27.6	21.7	66.9	55.1	43.3	120	98.4	77.3
10 × 10	90.25	53.7	44.2	34.7	107	88.4	69.5	192	158	124
12 × 12	132.25	78.7	64.8	50.9	157	130	102	281	231	182
14 × 14	182.25	108	89.3	70.2	217	179	140	387	319	251
16 × 16	240.25	143	118	92.5	286	235	185	511	420	330
18 × 18	306.25	182	150	118	364	300	236	651	536	421
20 × 20	380.25	226	186	146	452	373	293	808	665	523
22 × 22	462.25	275	227	178	550	453	356	982	809	636
24 × 24	552.25	329	271	213	657	541	425	1,174	966	759

Table 6.1 *continued*

Metric Units

Compression Strength, P (kN)

Column Size	A_g, mm^2	$F_c = 4{,}830\ kN/m^2$ C_P			$F_c = 9{,}650\ kN/m^2$ C_P			$F_c = 17{,}240\ kN/m^2$ C_P		
		0.85	0.70	0.55	0.85	0.70	0.55	0.85	0.70	0.55
5 × 5	13,064	53.6	44.1	34.7	107	88.3	69.4	191	158	124
6 × 6	19,516	80.1	65.9	51.8	160	132	104	286	235	185
8 × 8	36,290	149	123	96.3	298	245	193	532	438	344
10 × 10	58,226	239	197	155	478	393	309	853	703	552
12 × 12	85,322	350	288	226	700	577	453	1,250	1,029	809
14 × 14	117,580	482	397	312	965	794	624	1,723	1,419	1,115
16 × 16	155,000	636	524	411	1,272	1,047	823	2,271	1,870	1,469
18 × 18	197,580	811	668	524	1,621	1,335	1,049	2,895	2,384	1,873
20 × 20	245,322	1,006	829	651	2,013	1,658	1,302	3,594	2,960	2,326
22 × 22	298,225	1,223	1,008	792	2,447	2,015	1,583	4,369	3,598	2,827
24 × 24	356,290	1,462	1,204	946	2,923	2,407	1,892	5,220	4,299	3,378

Paul W. McMullin

where:

P = axial compression load, lb (kN)

A_g = gross cross-section area of column, in^2 (mm^2)

A_{brg} = gross cross-section area of bearing area, in^2 (mm^2)

Unlike for beams, axial stress is constant along the length of a column.

As long as the compressive stress is less than the adjusted design stress, we know we are OK. If it is higher, we select a larger column and recalculate things. Column design is iterative. We select a size, check it, and then resize as necessary.

When the column stability factor is close to or greater than 0.50, and the column bears on wood of the same species, the compression perpendicular to grain usually governs the column size. If cross-grain bearing controls, one may place the column on a steel plate to enlarge the bearing area, without making the column unnecessarily large.

6.3.1 Combined Compression and Bending

When a compression member carries compression and bending, the bending causes the stresses on one side of the member to increase, while decreasing stress on the opposite side, as illustrated in Figure 6.7. The maximum stress depends on the relative magnitudes of compression and bending. The increase from bending stress will reduce the axial load the member can carry. Combined stresses are common in truss top chords where they carry axial compression from truss action and bending from gravity loads.

To determine the combined effect of compression and bending stresses, we use the following unity equation. It is essentially the sum of the ratio of stress to adjusted design stress, added together. As long as the sum of these ratios is less than one, the member is OK.

$$\left(\frac{f_c}{F'_c}\right)^2 + \frac{f_b}{F'_b\left[1-\left(\frac{f_c}{F_{cE}}\right)\right]} \le 1.0$$

(6.10)

where:

f_c = axial stress, lb/in^2 (kN/m^2)

f_b = bending stress, lb/in^2 (kN/m^2)

F'_c = adjusted compression stress, lb/in^2 (kN/m^2)

F'_b = adjusted bending stress, lb/in^2 (kN/m^2)

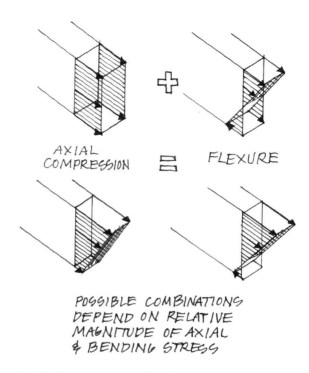

AXIAL COMPRESSION

FLEXURE

POSSIBLE COMBINATIONS
DEPEND ON RELATIVE
MAGNITUDE OF AXIAL
& BENDING STRESS

Figure 6.7 Combined compression and bending stresses

6.4 DEFLECTION

Column deformation is all axial—along its length. If a compression load is applied to our foam examples from before, we see the circles deform and become elliptical—shorter in the direction of compression—as shown in Figure 6.8a. If we combine axial load and bending, we see the circles on one side remain round, while they become more compressed on the other—see Figure 6.8b.

Axial shortening of columns is generally not a concern for the overall performance of a building. However, if you want to calculate it in your spare time, here's the equation:

$$\delta = \frac{Pl}{AE}$$

(6.11)

Paul W. McMullin

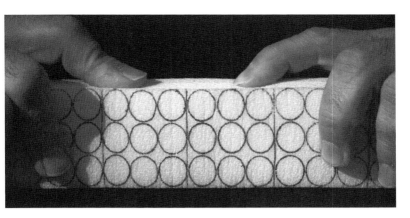

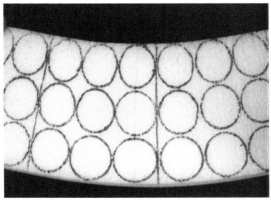

Figure 6.8 Foam column showing (a) pure compression and (b) combined compression and bending deformation

where:

P = axial force, lb (kN)

L = length, ft (m)

A = cross section area, in^2 (m^2)

E = modulus of elasticity, lb/in^2 (kN/m^2)

Remember to watch your units, so the numbers play nicely together. Also, refer to Chapter 4 for a discussion on long- and short-term deflection.

6.5 DETAILING

Column detailing is relatively simple. Some things to consider:

- Make sure the column ends are cut perpendicular to the length to ensure uniform bearing stresses.
- In a bolted connection, we ignore the holes when determining column area, as they are filled with the bolts, which can transmit force.
- It is not a very good idea to notch columns.

6.6 DESIGN STEPS

1. Draw the structural layout; include span dimensions and tributary width.
2. Determine loads:
   (a) unit loads;
   (b) load combinations yielding a point load;
   (c) member axial force.
3. Material parameters—find reference values and adjustment factors.
4. Estimate initial size.
5. Calculate stress and compare with adjusted design stress.
6. Summarize the results.

6.7 DESIGN EXAMPLE

Step 1: Draw Structural Layout

Figure 6.9 shows the structural layout. We will be designing the bottom-floor, interior column.

Step 2: Loads

Step 2a: Unit Loads

The unit loads are as follows:

$q_D = 20$ lb/ft^2	$q_D = 0.958$ kN/m^2
$q_L = 40$ lb/ft^2	$q_L = 1.92$ kN/m^2
$q_S = 55$ lb/ft^2	$q_S = 2.63$ kN/m^2

Paul W. McMullin

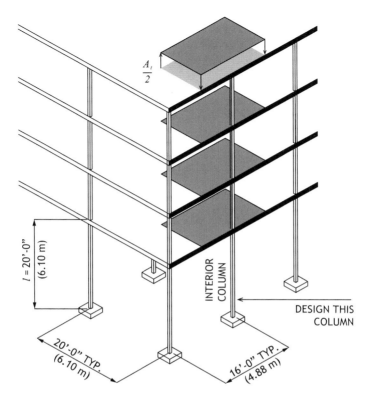

Figure 6.9 Column example layout

The tributary area for one floor is:

$$A_t = (16 \text{ ft})(20 \text{ ft}) = 320 \text{ ft}^2 \qquad A_t = (4.88 \text{ m})(6.10 \text{ m}) = 29.8 \text{ m}^2$$

Multiplying the unit load by the tributary area, we find the point load on each floor, for each load case:

$$P_D = q_D A_t$$

$$= 20 \frac{\text{lb}}{\text{ft}^2}(320 \text{ ft}^2)\frac{1 \text{ k}}{1000 \text{ lb}} = 6.4 \text{ k} \qquad = 0.958 \frac{\text{kN}}{\text{m}^2}(29.8 \text{ m}^2) = 28.5 \text{ kN}$$

Timber Compression   141

$$P_L = q_L A_t$$

$$= 0.040 \frac{k}{ft^2}(320 \text{ ft}^2) = 12.8 \text{ k} \qquad = 1.92 \frac{kN}{m^2}(29.8 \text{ m}^2) = 57.2 \text{ kN}$$

$$P_S = q_S A_t$$

$$= 0.055 \frac{k}{ft^2}(320 \text{ ft}^2) = 17.6 \text{ k} \qquad = 2.63 \frac{kN}{m^2}(29.8 \text{ m}^2) = 78.4 \text{ kN}$$

Step 2c: Load Combinations

We next combine these loads using load combinations. Because we don't know offhand which combination will control, we check the three that may. We also need to include the effect of each story. The multipliers on each load are the number of stories.

$P = 4P_D + 3P_L$ (Dead + Live Combination)
$= 4\ (6.4 \text{ k}) + 3\ (12.8 \text{ k}) \qquad = 4\ (28.5 \text{ kN}) + 3\ (57.2 \text{ kN})$
$= 64.0 \text{ k} \qquad\qquad\qquad\quad = 285.6 \text{ kN}$

$P = 4P_D + P_S$ (Dead + Snow Combination)
$= 4\ (6.4 \text{ k}) + 17.6 \text{ k} \qquad = 4\ (28.5 \text{ kN}) + 78.4 \text{ kN}$
$= 43.2 \text{ k} \qquad\qquad\qquad = 192 \text{ kN}$

$P = 4P_D + 3(0.75)P_L + 0.75P_S$ (Dead + 75% Live and Snow)
$= 4\ (6.4 \text{ k}) + 3\ (0.75)(12.8 \text{ k}) + 0.75\ (17.6 \text{ k}) = 67.6 \text{ k}$

$= 4\ (28.5 \text{ kN}) + 3\ (0.75)(57.2 \text{ kN}) + 0.75\ (78.4 \text{ kN}) = 301.5 \text{ kN}$

We see D + 0.75L + 0.75S controls, and so we will use $P = 67.6$ k (302 kN). We would not have known this without checking all three.

Drawing the free body diagram of the column, we get Figure 6.10.

Step 3: Material Parameters

We chose Southern Pine No 1. Key material parameters from Table A2.4 include compressive strength and minimum elastic modulus.

$$F_c = 1{,}500 \frac{lb}{in^2} \qquad\qquad F_c = 10{,}342 \frac{kN}{m^2}$$

$$E_{min} = 580 \frac{k}{in^2} \qquad\qquad E_{min} = 3{,}999 \frac{MN}{m^2}$$

Paul W. McMullin

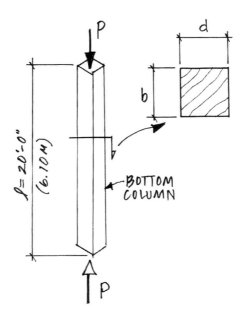

Figure 6.10 Column example free body diagram

Using Table 2.4, we find the adjustment factors that apply, specifically:

$F'_c = F_c C_D C_M C_t C_F C_i C_P$

$E'_{min} = E_{min} C_M C_t C_i C_T$

The easy adjustment factors are shown in the table.

Factor	Description	Source
$C_D = 1.0$	Load duration—normal	Table A4.1
$C_M = 1.0$	Wet service—dry	Table A4.2
$C_t = 1.0$	Temperature—normal	Table A4.4
$C_F = 1.0$	Size—factor built into reference design stress	Table A4.5
$C_i = 1.0$	Incising—not treated	Table A4.10
$C_T = 1.0$	Buckling stiffness—not a truss chord	Section 2.4.13

We have what we need to calculate E'_{min}:

$$E'_{min} = 580 \frac{k}{in^2}(1.0) = 580 \frac{k}{in^2} \qquad E'_{min} = 3{,}999 \frac{MN}{m^2}(1.0) = 3{,}999 \frac{MN}{m^2}$$

Step 4: Initial Size

We can guess an initial size of 12 in (300 mm) square. From Table A1.1, the cross-sectional area is $A_g = 132.3$ in² (58,081 mm²).

Or, we can calculate an initial size, assuming $C_p = 0.5$ (remember to calculate this for the final size).

$$A_{req} = \frac{P}{0.5F_c^*}$$

$$= \frac{67.6 \text{ k}}{0.5(1{,}500 \text{ lb/in}^2)} \frac{1000 \text{ lb}}{1 \text{ k}} \qquad = \frac{301.5 \text{ kN}}{0.5(10{,}342 \text{ kN/m}^2)}\left(\frac{1000 \text{ mm}}{1 \text{ m}}\right)^2$$

$$= 90.1 \text{ in}^2 \qquad\qquad\qquad = 58{,}306 \text{ mm}^2$$

Taking the square root of the area, we get 9.49 in (241 mm). Given this, let's try a 10 in (250 mm) square, solid sawn column. From Table A1.1, we write out the section properties we will need.

$A_g = 90.25$ in² $\qquad\qquad A_g = 58{,}081$ mm²

$b = d = 9.5$ in $\qquad\qquad b = d = 241$ mm

Step 5: Strength

With our assumed size, we can calculate C_p and adjust the reference design stress. Note the similarities with the beam stability equation.

Take your time and work through this one step at a time. Begin by writing out each variable and either the value or equation for it.

$$F_{CE} = \frac{0.822E'_{min}}{\left(\frac{l_e}{d}\right)^2} \qquad C_p = \frac{1+\left(F_{CE}/F_C^*\right)}{2C} \sqrt{\left[\frac{1+\left(F_{CE}/F_C^*\right)}{2C}\right]^2} - \frac{F_{CE}/F_C^*}{C}$$

$$l_e = l k_e$$

where:

l = unbraced length = 20 ft (6.1 m)

k_e = effective length factor = 1.0

$$l_e = 20 \text{ ft}(1.0)\frac{12 \text{ in}}{1 \text{ ft}} = 240 \text{ in} \qquad l_e = 6.1 \text{ m}(1.0)\left(\frac{1000 \text{ mm}}{1 \text{ m}}\right) = 6{,}100 \text{ mm}$$

Checking slenderness, we want to ensure:

$$\frac{l_e}{d} < 50$$

$$\frac{l_e}{d} = \frac{240 \text{ in}}{9.5 \text{ in}} = 25.3 \qquad\qquad \frac{l_e}{d} = \frac{6{,}100 \text{ mm}}{241 \text{ mm}} = 25.3$$

Because this is less than 50, the member is not slender, and we continue.

$$F_{CE} = \frac{0.822\left(580{,}000 \text{ lb/in}^2\right)}{(25.3)^2} \qquad\qquad F_{CE} = \frac{0.822\left(3{,}999{,}000 \text{ kN/m}^2\right)}{(25.3)^2}$$

$$= 745 \frac{\text{lb}}{\text{in}^2} \qquad\qquad\qquad\qquad = 5{,}136 \frac{\text{kN}}{\text{m}^2}$$

Recalling that F^*_c is F_c multiplied by all the adjustment factors but C_P:

$$F^*_c = F_c C_M C_t C_F C_i$$

$$= 1{,}500 \frac{\text{lb}}{\text{in}^2}(1.0) = 1{,}500 \frac{\text{lb}}{\text{in}^2} \qquad\qquad = 10{,}342 \frac{\text{kN}}{\text{m}^2}(1.0) = 10{,}342 \frac{\text{kN}}{\text{m}^2}$$

$c = 0.8$ (for sawn lumber)

We now have what we need to calculate the column stability factor.

$$C_P = \frac{1 + \left(745/1{,}500\right)}{2(0.8)} - \sqrt{\left[\frac{1 + \left(745/1{,}500\right)}{2(0.8)}\right]^2 - \frac{745/1{,}500}{0.8}}$$

$$= 0.43$$

$$C_P = \frac{1 + \left(5{,}136/10{,}342\right)}{2(0.8)} - \sqrt{\left[\frac{1 + \left(5{,}136/10{,}342\right)}{2(0.8)}\right]^2 - \frac{5{,}136/10{,}342}{0.8}}$$

$$= 0.43$$

And now, the adjusted design stress is:

$$F'_c = F^*_c C_P$$

$$= 1{,}500 \frac{\text{lb}}{\text{in}^2}(0.43) = 645 \frac{\text{lb}}{\text{in}^2} \qquad\qquad = 10{,}342 \frac{\text{kN}}{\text{mm}^2}(0.43) = 4{,}447 \frac{\text{kN}}{\text{mm}^2}$$

Calculating the column stress, we find:

$$f_c = \frac{P_u}{A_g}$$

$$= \frac{67.6\ \text{k}}{90.25\ \text{in}^2}\frac{1000\ \text{lb}}{1\ \text{k}} = 749\frac{\text{lb}}{\text{in}^2} \qquad = \frac{301.5\ \text{kN}}{58{,}081\ \text{mm}^2}\left(\frac{1000\ \text{mm}}{1\ \text{m}}\right)^2 = 5{,}191\frac{\text{kN}}{\text{m}^2}$$

Unfortunately, f_c is greater than F'_c, and so we need to go up a size. With columns, we have two things working against us. As the section size gets smaller, the stress goes up. At the same time, the buckling stability factor goes down, reducing our allowable stress.

Trying a 12in (305 mm) square column:

$$A_g = 132.3\ \text{in}^2 \qquad\qquad A_g = 85{,}264\ \text{mm}^2$$

$$f_c = \frac{67.6\ \text{k}}{132.3\ \text{in}^2}\frac{1000\ \text{lb}}{1\ \text{k}} = 511\frac{\text{lb}}{\text{in}^2} \qquad = \frac{301.5\ \text{kN}}{85{,}264\ \text{mm}^2}\left(\frac{1000\ \text{mm}}{1\ \text{m}}\right)^2 = 3{,}536\frac{\text{kN}}{\text{m}^2}$$

Without recalculating C_P, we see the compressive stress is below the allowable level for the 10 in (250 mm) column. Perhaps you can recalculate C_P for the 12 in (305 mm) column.

Step 6: Summary

In summary, we have a 12×12 Southern Pine No. 1 column.

6.8 WHERE WE GO FROM HERE

We really don't have much further to go from here. We have looked at compression loads along the length of the member, perpendicular to the grain in a bearing condition, and when compression loads are combined with bending. From here, we practice!

NOTE

1. ANSI/AWC. *National Design Specification (NDS) Supplement:* *Design Values for Wood Construction* (Leesburg, VA: AWC, 2015).

Timber Trusses

Chapter 7

Frank P. Potter

Trusses have been used in structures for thousands of years because of their structural efficiency. They can be configured in limitless ways and span great distances. Today, trusses are commonly used in residential homes, warehouses, office buildings, hotels, apartment buildings, and industrial structures. They have aesthetic value that can be incorporated into the finished look of the structure.

7.1 STABILITY

A key concept in understanding trusses is the triangle—the most stable geometric shape. Imagine a fractionless pin at each end of a triangle connected by solid, rigid bars. If the triangle is pushed with a horizontal force, F, the triangle will hold its shape (see Figure 7.1). By way of comparison, now imagine a square or rectangle with the same pins and bars. The shape collapses when the same force is applied, regardless of magnitude, as illustrated in Figure 7.2.

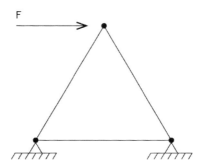

Figure 7.1 Basic truss element, the triangle

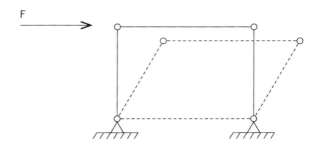

Figure 7.2 Rectangular frame with pinned member ends

Frank P. Potter

Rectangular frames are commonly used to resist horizontal applied loads in many structures. What inhibits this frame from falling? We lock the pins, as illustrated in Figure 7.3. These locked joints are more commonly known as moment connections, because they resist rotational forces (moments).

A truss, then, can be thought of as a large beam comprising a combination of triangles. This 'beam' resists applied loads spanning a distance. There are an infinite variety of triangle configurations that can be used to resist loads. A few are illustrated in Figure 7.4.

The analysis of a truss is handled in two separate phases, used to determine external and internal forces. Phase 1 is the global analysis. First, we determine applied loads acting on the truss (usually along the top chord in wood trusses). Second, we calculate the reactions. Just like beams, determinacy is critical at this point. The number of unknown reactions must be equal to the known statics equations. If this is not the case, more sophisticated analysis methods are required. Refer to Chapter 9 of *Introduction to Structures* in this series.[1]

Next, internal analysis looks at how the external forces (determined in phase 1) flow though the truss and determines the stresses in each member.

We must look at trusses as 3-D systems. Even if they appear to be planar elements, they must be braced out of plane to keep them stable. With timber trusses, the entire top chord will be in compression and bending, as described above. If plywood roof sheathing or bridging is

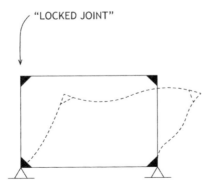

"LOCKED JOINT"

Figure 7.3 Rectangular frame with fixed joints (**moment frame**)

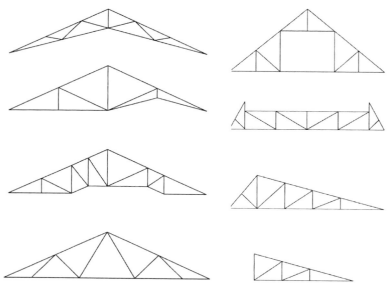

Figure 7.4 Various truss configuations

applied, the capacity of the truss to support loads is greatly increased. Without the plywood, the truss has a tendency to roll over.

7.2 CAPACITY

The beauty of trusses is that the majority of members only have axial loads. Each member will either be in tension or compression. Consider the simple truss in Figure 7.5: you can see how a horizontal force, *F*, applied in either direction left to right resolves into only tension or compression forces. This is why trusses are so efficient.

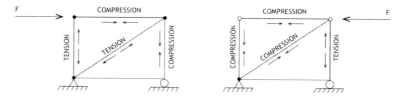

Figure 7.5 Horizontal force resolution in a truss

　　　　　Frank P. Potter

The only exception is where the truss experiences applied loads between joints on either the top or bottom chords. These loads are usually distributed or point loads. Figure 7.6a illustrates the effect of these loads on the member forces. By way of example, imagine a common wood truss in the roof of a home. The top chord supports the roof's dead, snow, or rain loads. The chord member between the joints experiences bending, but distributes the bending loads as axial forces to the remaining truss members, shown in Figure 7.6b. The bending forces substantially increase the size of the truss member.

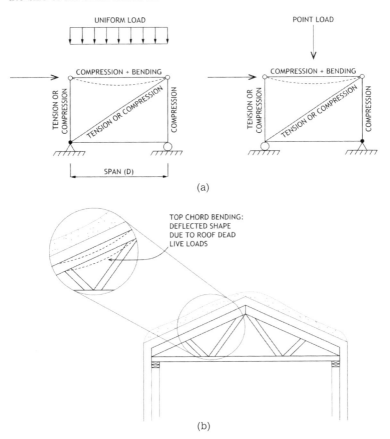

Figure 7.6 (a) Combined axial and bending loads in a truss, and (b) bending deflection in a roof truss

7.3 TRUSS ANALYSIS

A large part of the analysis of trusses comes from the derivation of the external loads and how to apply them according to the building and material codes. This was discussed in detail in *Introduction to Structures*. The one thing to keep in mind for this text is that all design will be per allowable stress provisions versus ultimate strength.

Continuing with the roof truss previously mentioned, the governing building codes prescribe what loads and percentages of those loads should be applied to the truss. These are known as load combinations. Examples of these are as follows:

D
$D + L$
$D + (L_r$ or S or $R)$
$D + 0.75L + .75(L_r$ or S or $R)$.

where:

D = dead load;
L = live load (usually relating to people);
L_r = roof live load (minimum prescribed loads in areas without snow);
S = snow load;
R = rain loads.

Once we determine loads and apply them to the appropriate location (i.e., the truss top chord), we calculate the external reactions at the supports (i.e., global analysis) using statics. From that point, the internal analysis can determine the force in each member and then design wood sections to resist them.

The commonest methods for analyzing wood trusses by hand are the method of joints and the method of sections. Both are discussed in great detail in Chapter 10 of *Introduction to Structures*. They begin with global analysis, defining the geometry, loads, and support reactions.

7.3.1 Method of Joints

The method of joints focuses on forces that occur at each joint of the truss. Each joint theoretically acts as a frictionless hinge, transferring no moment. Another way to think of it is that the truss members are loosely bolted together and, therefore, free to rotate.

Say we are analyzing the truss in Figure 7.7. Start with a joint where you have knowledge of the external forces—initially at the supports. Draw a free body diagram of the forces acting on the joint (parallel to the direction of the members), illustrated in Figure 7.8. Sum the horizontal and vertical forces to solve for the unknown member forces. Once all of the forces have been calculated, simply move to the next adjacent joint. Set up another free body diagram and repeat the procedure. Remember to break diagonal forces down into their horizontal and vertical components.

7.3.2 Method of Sections

The previous method can be quite laborious, especially if you only need forces in a few members. The method of sections drastically minimizes the effort and does not require analysis of the entire truss. We strategically slice the truss into two pieces, through the members that are under investigation, shown in Figure 7.9. At the cut, we place axial forces parallel

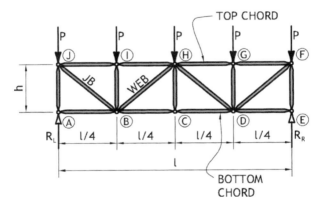

Figure 7.7 Truss layout and loading

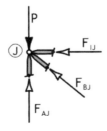

Figure 7.8 Joint free body diagram for method of joints

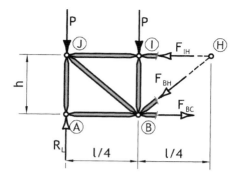

Figure 7.9 Truss segment free body diagram for method of sections

to the members, and then sum moments where the multiple cut members intersect. This is similar to the initial global analysis. We finish by summing vertical and horizontal forces to determine any remaining unknown forces in the members under consideration.

7.4 DEMAND VS. CAPACITY

Now, we size members using the provisions for calculating tension, bending, and compression stress that we discussed in Chapters 3, 4, and 6, respectively. You already have the tools to design truss members!

7.5 DEFLECTION

One final factor to consider when designing trusses is the deflection. Deflection limits for the trusses are the same as for other structural members. The governing building codes prescribe minimum limits for the allowable deflection under dead and live loads (see Section 4.4). For a typical floor truss, the total load deflection is not to exceed the total truss span (in inches or millimeters) divided by 240. If the truss spans 30 ft (9.4 m), the allowable code deflection is:

$$\delta_{a.TOT} = \frac{l}{240}$$

$$= \frac{30 \text{ ft}\left(12 \text{ in}/1 \text{ ft}\right)}{240} = 1.5 \text{ in} \qquad\qquad = \frac{9,140 \text{ mm}}{240} = 38 \text{ mm}$$

Frank P. Potter

Truss deflection is a function of span length, loads, and modulus of elasticity. The latter varies, based on the wood species and grade. We use stiffness or energy methods to determine truss deflections. These are a bit more involved and beyond the scope of this book. However, we can make a quick approximation of deflection using the beam deflection equations in Appendix 6. We can estimate the moment of inertia of the truss as:

$$I = 0.75Ad^2 \tag{7.1}$$

where:

A = the cross-sectional area of one chord, in^2 (mm^2)

d = distance between the chord centroid, in (mm)

7.6 DETAILING

With the evolution of computers able to perform sophisticated truss analysis, the shapes of trusses and their function are unlimited. Figure 7.4 shows various shapes that can be used, not only to support the roof dead and live loads, but also to create architectural effects inside and outside the structure.

The top chord of the truss can have multiple slopes and pitches, not only to efficiently shed snow and rain, but also to dramatically impact the look of the structure. The bottom chord can also be manipulated to vaulted spaces at any point along the span. If a truss is deep enough, a room (attic space) can be incorporated into the design.

Truss chords and webs of trusses can be connected together in many different ways. Trusses made with dimensional lumber (i.e., 2 × 4, 2 × 6, etc.) are commonly connected together with metal plates, also called gusset or gang nail plates. These gusset plates are made with sheet metal of varying thickness, often with little teeth. These kinds of connection are intended for rapid production and are usually found in residential construction. Computer programs determine the lengths and cut angles of each truss member. These are laid out on a table, and the gusset plates are pressed in by hydraulic rollers. These light trusses can be assembled in minutes.

Where trusses are intended to be seen inside the structure, it is common practice for architects to use heavy timbers and express the connection. The connections of the chords and webs are comprised of ¼ in (6.35 mm) steel plates and 1 in (25.4 mm) diameter and larger bolts. These plates can be hidden or exposed, as illustrated in Figure 7.10. Hidden plates are

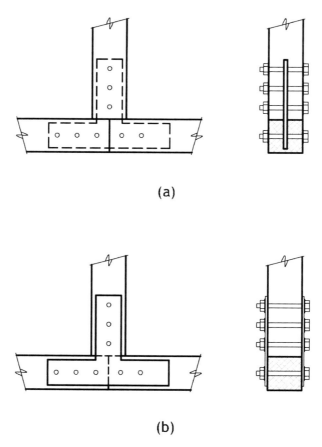

Figure 7.10 (a) Hidden and (b) exposed connectors in a heavy timber truss

commonly referrred to as knife plates. The truss member is sliced in the middle (usually with a chain saw, which is the perfect width for this kind of connection), and the steel plate is slipped in the cut, and then bolts are inserted through.

Exposed plates are applied on the outside of the truss. In this case, two plates are required at each connection. Exposed plates require fewer bolts, as they are in double shear, as shown in Figure 7.10b.

Frank P. Potter

7.7 DESIGN STEPS

The steps for truss design are as follows:

1. Determine the truss profile and spacing that best suit the structural conditions.
   (a) Label each joint and member in the truss profile.
2. Determine the loads and forces:
   (a) unit loads;
   (b) load combinations yielding line or point loads;
   (c) perform global analysis to determine truss support reactions;
   (d) perform internal analysis to determine member forces (method of joints or method of sections).
3. Material parameters—find the reference design values and adjustment factors.
4. Estimate initial size of the members.
5. Calculate member stresses and compare with the adjusted design values:
   (a) axial stress (tension and compression);
   (b) bending and combined stress.
6. Calculate deflection and allowable deflections and compare them.
7. Summarize the results.

7.8 DESIGN EXAMPLE

Step 1: Determine the Structural Layout

Consider the simple parallel top and bottom chord truss shown in Figure 7.11. This style of truss is common for floor and roof levels in high-density residential structures. The trusses are spaced at 4 ft (1.2 m) on center.

Step 2: Determine the Loads and Member Forces

Step 2a: Unit Loads

The unit loads are:

$D = 25$ lb/ft^2 (1.20 kN/m^2)

$L = 25$ lb/ft^2 (1.20 kN/m^2)

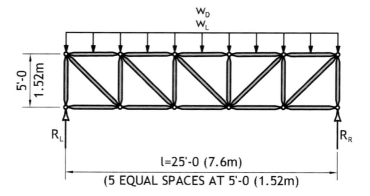

Figure 7.11 Example truss layout and loads

Moving forward, this example will analyze the truss with 100 percent of the total dead load and live load. The uniform load on the top chord is:

$$w = (D + L)l_t$$

$$= \left(25\frac{\text{lb}}{\text{ft}^2} + 25\frac{\text{lb}}{\text{ft}^2}\right)4 \text{ ft}$$

$$= 200\frac{\text{lb}}{\text{ft}}$$

$$= \left(1.20\frac{\text{kN}}{\text{m}^2} + 1.20\frac{\text{kN}}{\text{m}^2}\right)1.22 \text{ m}$$

$$= 2.93\frac{\text{kN}}{\text{m}}$$

Step 2c: Perform Global Truss Analysis

The next step is to label all joints and members of the truss. Organization is key to truss analysis. We will label joints as numbers and members as letters, as shown in Figure 7.12.

The global analysis can now begin. We start by finding the support reactions. From the diagram, the truss is supported by a pin connection at one end and a roller at the other. There are no horizontal reactions.

Because the load is uniform, we can find the reactions as the uniform load multiplied by the truss length divided by two (half of the load is supported at each support). In equation form:

$$R = \frac{wl}{2}$$

Frank P. Potter

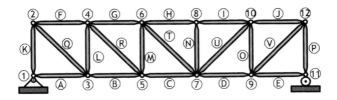

Figure 7.12 Example truss member and joint labels

$$= \frac{200 \text{ lb/ft (25 ft)}}{2}$$

$$= 2,500 \text{ lb}$$

$$= \frac{2.93 \text{ kN/m (7.6 m)}}{2}$$

$$= 11.13 \text{ kN}$$

Step 2d: Perform Internal Analysis

Our analysis can now shift internally. The focus is to determine the force in the webs and chords. We will find the forces in members Q, C, and H in this example.

Using the method of joints, we determine the force in web member Q.

Figure 7.13 shows the isolated joint 2. Force direction is important here. Forces acting upward and to the right will be positive, and forces acting downward and to the left will be negative. Looking back at the global analysis, the reaction at joint 1 is upwards (positive). Because member A is horizontal, it carries no vertical force. Therefore, member K carries all of the support reaction. This tells us the force is:

$K = 2,500$ lb (11.13 kN)

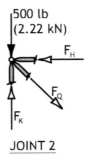

500 lb
(2.22 kN)

F_H

F_Q

F_K

JOINT 2

Figure 7.13 Example joint C for method of joints analysis

Timber Trusses

Moving to member Q, we assume the force acts away from the joint, but at a diagonal. This is tricky to analyze, and so we break it into horizontal and vertical components. The hypotenuse is 7.071 ft (2.16 m), and the vertical leg is 5 ft (1.52 m). To determine the force, use the static concept that the sum of all the forces in the Y direction have to equal zero:

$$\sum F_y = 0$$

$$2{,}500 \text{ lb} - F_Q\left(\frac{5 \text{ ft}}{7.07 \text{ ft}}\right) - 500 \text{ lb}$$

$$= 0$$

$$11.13 \text{ kN} - F_Q\left(\frac{1.52 \text{ m}}{2.16 \text{ m}}\right) - 2.22 \text{ kN}$$

$$= 0$$

Solving for FQ, we get:

$$F_Q = \frac{2{,}500 \text{ lb} - 500 \text{ lb}}{\left(5 \text{ ft}\middle/7.07 \text{ ft}\right)}$$

$$= 2{,}828 \text{ lb}$$

$$F_Q = \frac{11.13 \text{ kN} - 2.22 \text{ kN}}{\left(1.52 \text{ m}\middle/2.16 \text{ m}\right)}$$

$$= 12.7 \text{ kN}$$

Notice that the answers came out positive. This indicates that the assumptions of the force directions are correct. If the answers come out negative, then the forces are acting in the opposite direction. In this case, the web member is in tension.

Moving on to the forces in the top and bottom chords, H and C, we use the method of sections. We cut an imaginary section at the mid-span where the chord forces will be highest, as shown in Figure 7.14. The section is cut through the two members in question. This method is based on the static concept of summing moments about a point. In determining forces on member H, the focus will be on joint 7, even though it is outside the truss.

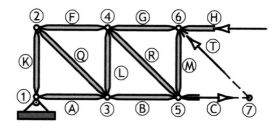

Figure 7.14 Example free body diagram for method of sections

Frank P. Potter

The external forces acting on the truss are the uniform load on the top chord and the reactions at the supports. **Sign convention** again is important here. Understand that a moment is a rotational force, and that to determine a moment you multiply the force by the perpendicular distance; moments going in a counterclockwise (left) direction will be positive, and moments going clockwise (right) will be negative. If the moments around joint 7 are summed, moments from members H and T drop out, because their **moment arm** length is 0, leaving:

$$\sum M_7 = 0$$

$$w(15 \text{ ft})\frac{15 \text{ ft}}{2} - F_K(15 \text{ ft}) + F_H(5 \text{ ft}) = 0$$

$$200\frac{\text{lb}}{\text{ft}}\frac{(15 \text{ ft})^2}{2} - 2{,}500 \text{ lb } (15 \text{ ft}) + F_H(5 \text{ ft}) = 0$$

Rearranging

$$F_H = \frac{-200 \text{ lb/ft} \frac{(15 \text{ ft})^2}{2} + 2{,}500 \text{ lb } (15 \text{ ft})}{5 \text{ ft}}$$

$$= 3{,}000 \text{ lb}$$

$$w(4.56 \text{ m})\frac{4.56 \text{ m}}{2} - F_K(4.56 \text{ m}) + F_H(1.52 \text{ m}) = 0$$

$$-2.93\frac{\text{kN}}{\text{m}}\frac{(4.56 \text{ m})^2}{2} - 11.13 \text{ kN } (4.56 \text{ m}) + F_H(1.52 \text{ m}) = 0$$

Rearranging

$$F_H = \frac{-2.93 \text{ kN/m} \frac{(4.56 \text{ m})^2}{2} + 11.13 \text{ kN } (4.56 \text{ m})}{1.52 \text{ m}}$$

$$= 13.35 \text{ kN}$$

Now, to find the force in member C, sum the moment around joint 6 (refer to Figure 7.14). It turns out that F_C is of the same magnitude as F_H, just in a different direction. This makes sense, given the symmetrical loading and need for static equilibrium. Consider running through the calculations yourself to confirm this.

Step 3: Material Parameters

We will need adjusted design stresses for tension and compression. We will use Douglas Fir–Larch No. 2, as this is available in the project location.

Reference design stresses from Table A2.1 are as follows:

$$F_t = 500 \frac{\text{lb}}{\text{in}^2} \qquad\qquad F_t = 3{,}447 \frac{\text{kN}}{\text{m}^2}$$

$$F_c = 1{,}400 \frac{\text{lb}}{\text{in}^2} \qquad\qquad F_c = 9{,}653 \frac{\text{kN}}{\text{m}^2}$$

$$E_{\min} = 470{,}000 \frac{\text{lb}}{\text{in}^2} \qquad\qquad E_{\min} = 3{,}240{,}535 \frac{\text{kN}}{\text{m}^2}$$

Using Table 2.4, we find the adjustment factors that apply for tension and compression, specifically:

$$F'_t = F_t C_D C_M C_t C_F C_i$$
$$F'_c = F_c C_D C_M C_t C_F C_i C_P$$
$$E'_{\min} = E_{\min} C_M C_t C_i C_T$$

The easy adjustment factors are shown in the table.

Factor	Description	Source
$C_D = 1.0$	Load duration—normal	Table A4.1
$C_M = 1.0$	Wet service—dry	Table A4.2
$C_t = 1.0$	Temperature—normal	Table A4.4
$C_F = 1.3$	Size—tension (2 × 4)	Table A4.5
$C_F = 1.2$	Size—bending (2 × 8)	Table A4.5
$C_F = 1.05$	Size—compression (2 × 8)	Table A4.5
$C_i = 1.0$	Incising—not treated	Table A4.10
$C_T = 1.15$	Buckling stiffness—provided for convenience	Section 2.4.13

Let's calculate F'_t and $E'_{\min}$ now:

$$F'_t = 500 \frac{\text{lb}}{\text{in}^2}(1.0)1.3 \qquad\qquad F'_t = 3{,}447 \frac{\text{kN}}{\text{m}^2}(1.0)1.3$$

$$= 650 \frac{\text{lb}}{\text{in}^2} \qquad\qquad\qquad = 4{,}481 \frac{\text{kN}}{\text{m}^2}$$

$$E'_{\min} = 470{,}000 \frac{\text{lb}}{\text{in}^2}(1.0)1.15 \qquad E'_{\min} = 3{,}240{,}535 \frac{\text{kN}}{\text{m}^2}(1.0)1.15$$

$$= 540{,}500 \frac{\text{lb}}{\text{in}^2} \qquad\qquad\qquad = 3{,}726{,}615 \frac{\text{kN}}{\text{m}^2}$$

Frank P. Potter

Step 4: Initial Size

For simplicity, let's make the webs and bottom chords 2 × 4 (50 × 102), and the top chord 2 × 10 (50 × 250) to help with the bending.

Step 5: Calculate Member Stresses

Step 5a: Axial Stress

The top chord will be the starting member. Recall from the analysis that this member is in compression. In addition, it will also have bending owing to the roof dead and live loads. So, this member will have the unique condition of experiencing combined stresses. We follow the steps of Chapter 6 for determining axial stress.

We first need the column stability factor. As the weak axis is braced by the sheathing, we look at the strong axis, $d = 7.25$ in (184 mm). This yields:

$$C_P = \frac{1 + \left(\dfrac{F_{CE}}{F_C^*}\right)}{2c} - \sqrt{\left[\frac{1 + \left(\dfrac{F_{CE}}{F_C^*}\right)}{2c}\right]^2 - \frac{\left(\dfrac{F_{CE}}{F_C^*}\right)}{c}}$$

$$F_{CE} = \frac{0.822 E'_{min}}{\left(\dfrac{l_e}{d}\right)^2}$$

$$l_e = l k_e$$

where:

l = unbraced length = 5 ft (1.52 m)

k_e = effective length factor = 1.0

$$l_e = 5 \text{ ft}(1.0)\frac{12 \text{ in}}{1 \text{ ft}} = 60 \text{ in} \qquad l_e = 1.52 \text{ m}(1.0)\frac{1000 \text{ mm}}{1 \text{ m}} = 1{,}520 \text{ mm}$$

Checking slenderness, we want to ensure that:

$$\frac{l_e}{d} < 50$$

$$\frac{l_e}{d} = \frac{60 \text{ in}}{7.25 \text{ in}} = 8.3 \qquad \frac{l_e}{d} = \frac{1{,}520 \text{ mm}}{184 \text{ mm}} = 8.3$$

Because this is less than 50, the member is not slender, and we continue:

$$F_{CE} = \frac{0.822\left(540{,}500 \text{ lb/in}^2\right)}{(8.3)^2} = 6{,}449 \frac{\text{lb}}{\text{in}^2}$$

$$F_{CE} = \frac{0.822\left(3{,}726{,}535 \text{ kN/m}^2\right)}{(8.3)^2} = 44{,}466 \frac{\text{kN}}{\text{m}^2}$$

Recalling that F^*_c is F_c multiplied by all the adjustment factors but C_P:

$$F^*_c = F_c C_M C_t C_F C_i$$

$$= 1{,}400 \frac{\text{lb}}{\text{in}^2}(1.0)1.05 = 1{,}470 \frac{\text{lb}}{\text{in}^2} \qquad = 9{,}653 \frac{\text{kN}}{\text{m}^2}(1.0)1.05 = 10{,}136 \frac{\text{kN}}{\text{m}^2}$$

$c = 0.8$ (for sawn lumber)

We now have what we need to calculate the column stability factor.

$$C_P = \frac{1+\left(6{,}449 \big/ 1{,}470\right)}{2(0.8)} - \sqrt{\left[\frac{1+\left(6{,}449 \big/ 1{,}470\right)}{2(0.8)}\right]^2 - \frac{6{,}449 \big/ 1{,}470}{0.8}}$$

$$= 0.95$$

$$C_P = \frac{1+\left(44{,}466 \big/ 10{,}136\right)}{2(0.8)} - \sqrt{\left[\frac{1+\left(44{,}466 \big/ 10{,}136\right)}{2(0.8)}\right]^2 - \frac{44{,}466 \big/ 10{,}136}{0.8}}$$

$$= 0.95$$

And now, the adjusted design stress is:

$$F'_c = F^*_c C_P$$

$$= 1{,}470 \frac{\text{lb}}{\text{in}^2}(0.95) = 1{,}397 \frac{\text{lb}}{\text{in}^2} \qquad = 10{,}136 \frac{\text{kN}}{\text{mm}^2}(0.95) = 9{,}629 \frac{\text{kN}}{\text{mm}^2}$$

Finding the axial stress in the top chord, we get:

$$f_c = \frac{F}{A}$$

$$= \frac{3{,}000 \text{ lb}}{1.5 \text{ in}(7.25 \text{ in})} \qquad = \frac{13.35 \text{ kN}}{38 \text{ mm }(184 \text{ mm})\left(1 \text{ m}\big/ 1000 \text{ mm}\right)^2}$$

$$= 276 \frac{\text{lb}}{\text{in}^2} \qquad\qquad = 1{,}909 \frac{\text{kN}}{\text{m}^2}$$

The member stress, f_c, is less than the adjusted design stress, F'_c, which is a good sign. However, we need to include the effect of the bending stress, which may have a substantial effect.

Finding the adjusted design stress

$$F'_b = F_b C_D C_M C_t C_L C_F C_{fu} C_i C_r$$

$$= 725 \frac{\text{lb}}{\text{in}^2}(1.0)1.2 = 870\frac{\text{lb}}{\text{in}^2} \qquad = 4,999\frac{\text{kN}}{\text{m}^2}(1.0)1.2 = 5,999\frac{\text{kN}}{\text{m}^2}$$

The moment in the top chord equals:

$$M = \frac{wl^2}{8}$$

$$= \frac{200 \text{ lb/ft (5 ft)}^2}{8} \qquad = \frac{293 \text{ kN/m (1.52 m)}^2}{8}$$
$$= 625 \text{ lb} - \text{ft} \qquad = 0.846 \text{ kN} - \text{m}$$

From Table A1.1, the section modulus for a 2 × 8 about the strong (X) axis is:

$$S_x = 13.14 \text{ in}^3 \qquad (0.214 \times 10^6 \text{ mm}^3)$$

$$f_b = \frac{625 \text{ lb} - \text{ft}}{13.14 \text{ in}^3}\left(\frac{12 \text{ in}}{1 \text{ ft}}\right) \qquad f_b = \frac{0.845 \text{ kN} - \text{m}}{0.214 \times 10^6 \text{ mm}^3\left(1 \text{ m}/1000 \text{ mm}\right)^3}$$

$$= 571\frac{\text{lb}}{\text{in}^2} \qquad = 3,949\frac{\text{kN}}{\text{m}^2}$$

Combining bending and compression stress, we have:

$$\left[\frac{f_c}{F'_c}\right]^2 + \frac{f_b}{F'_b\left[1-\left(f_c/F_{cE}\right)\right]} \leq 1.0$$

We have all the variables, we just need to plug them in.

$$\left[\frac{276}{1,397}\right]^2 + \frac{571}{870\left[1-\left(571_c/6,449\right)\right]} \qquad \left[\frac{1,909}{9,629}\right]^2 + \frac{3,949}{5,999\left[1-\left(3,947/44,466\right)\right]}$$

$$= 0.72 \qquad = 0.72$$

Because our ratio is less than one, our top chord works.

We now turn our attention to the other two members from the analysis, the diagonal web and the bottom chord. Interestingly, both members are in tension, and the forces are about the same magnitude. We will design for the higher of the two, or $F_Q = 3,000$ lb (13.4 kN).

Looking at member Q and recalling that the webs and bottom chords are initially 2 × 4s (50 × 102), we calculate the actual stresses:

$$f_t = \frac{F}{A}$$

$$= \frac{3,000 \text{ lb}}{1.5 \text{ in } (3.5 \text{ in})}$$

$$= 571 \frac{\text{lb}}{\text{in}^2}$$

$$= \frac{13.4 \text{ kN}}{38 \text{ mm } (89 \text{ mm})\left(1 \text{ m}/1000 \text{ mm}\right)^2}$$

$$= 3,947 \frac{\text{kN}}{\text{m}^2}$$

As this is less than our adjusted design stress, our members work at the size we selected.

$$F_t' = 500 \frac{\text{lb}}{\text{in}^2}(1.0)1.5$$

$$= 750 \frac{\text{lb}}{\text{in}^2}$$

$$F_t' = 3,447 \frac{\text{kN}}{\text{m}^2}(1.0)1.5$$

$$= 5,171 \frac{\text{kN}}{\text{m}^2}$$

This is greater than the tension stress. We know the bottom chord and last diagonal work for tension.

Step 7: Summarize Results

Because the diagonal members further towards the middle of the truss will have a lower tension force than the end one, we can use a 2 × 4 for all members of the truss.

7.9 WHERE WE GO FROM HERE

You now have the tools to design truss members made of light or heavy timber. From here, we can explore other truss configurations and get into the connection design.

NOTE

1. Paul W. McMullin and Jonathan S. Price. *Introduction to Structures* (New York: Routledge, 2016).

Timber Lateral Design

Chapter 8

Paul W. McMullin

8.1 INTRODUCTION

Lateral loads on structures are commonly caused by wind, earthquakes, and soil **pressure** and less commonly caused by human activity, waves, or blasts. These loads are difficult to quantify with any degree of precision. However, following reasonable member and system proportioning requirements, coupled with prudent detailing, we can build reliable timber structures that effectively resist lateral loads.

What makes a structure perform well in a windstorm is vastly different than in an earthquake. A heavy, squat structure, such as the Parthenon in Greece, can easily withstand wind—even without a roof. Its mass anchors it to the ground. At the other extreme, a tent structure could blow away in a moderate storm. Conversely, the mass of the Parthenon makes it extremely susceptible to earthquakes (remember earthquake force is a function of weight), whereas the tent, in a seismic event, will hardly notice what is going on.

Looking at this more closely, wind forces are dependent on three main variables:

1. proximity to open spaces such as water or mud flats;
2. site exposure;
3. building shape and height.

In contrast, earthquake forces are dependent on very different variables:

1. nature of the seismic event;
2. building weight;
3. rigidity of the structural system.

Because we operate in a world with gravity forces, we inherently understand the gravity load paths of the simple building shown in Figure 8.1a. Downward loads enter the roof and floors and make their way to the walls, columns, and, eventually, **footings**. Lateral loads can take more time to grasp. But, we can think of them as turning everything 90°, the structure acting as a cantilevered beam off the ground, as illustrated in Figure 8.1b.

The magnitude and distribution of lateral loads drive the layout of frames and **shear walls**. These walls resist lateral forces, acting like cantilevered beams poking out of the ground.

We design lateral wind-resisting members not to damage the system. Conversely, because strong seismic loads occur much less frequently, we design their lateral systems to crush the wood and yield connections.

Paul W. McMullin

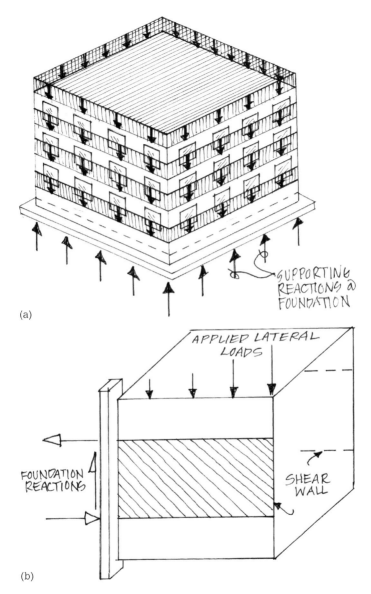

(a)

SUPPORTING
REACTIONS @
FOUNDATION

APPLIED LATERAL
LOADS

FOUNDATION
REACTIONS

SHEAR
WALL

(b)

Figure 8.1 (a) Gravity load path; (b) lateral load path turned 90°

Timber Lateral Design

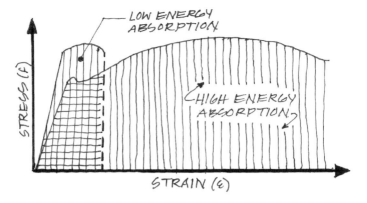

Figure 8.2 Comparative energy absorption for high and low deformation behavior

This absorbs significantly more energy, as illustrated in Figure 8.2, resulting in smaller member sizes. However, it leaves the structure damaged.

For the design of **seismic load resisting systems**, we follow rigorous member proportioning and detailing requirements to ensure crushing and yielding occur in the right places. This chapter focuses on design and detailing requirements from a conceptual point of view, and what lateral load resisting systems, elements, and connections should look like.

8.2 LATERAL LOAD PATHS

Following the path lateral loads travel through a structure is key to logical structural configuration and detailing. If the load path is not continuous from the roof to ground, failure can occur. Additionally, no amount of structural engineering can compensate for an unnecessarily complex load path.

When configuring the structure, visualize how lateral forces—and gravity forces—travel from element to element, and eventually to the ground. A well-planned load path will save weeks of design effort, substantially reduce construction cost, and minimize structural risk. Software can't do this, but careful thought will.

Looking at lateral load paths further, Figure 8.3 shows how they enter a structure and find their way to the ground. Starting at point 1, wind

Paul W. McMullin

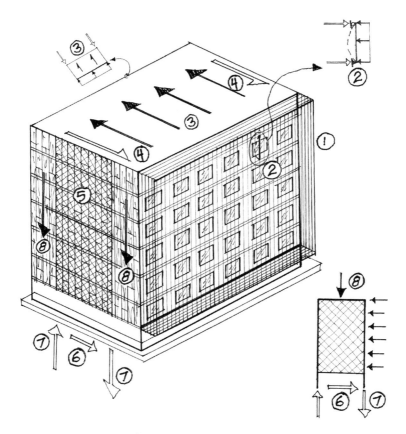

Figure 8.3 Detailed lateral load path in structure

induces pressure, or seismic accelerations cause inertial forces, perpendicular to the face of the building. Spanning vertically (point 2), the wall delivers a line or point load to a connection at the roof or floor level. The roof or floor picks up additional inertial seismic load. The roof (number 3) must resist lateral forces through diaphragm action— essentially a deep beam. The ends of the diaphragm (point 4) then deliver load into connections between a shear wall or frame. This occurs at each level (point 5). The lateral force works its way to the footing (point 6), which transfers the force to the soil through friction and passive pressure.

Because the lateral forces are applied at a distance above the ground, they impart an overturning moment to the system. This causes tension and compression in the ends of shear walls and outside frame columns (point 7). The weight of the structure (point 8) helps resist this overturning moment, keeping it from tipping over.

To review, lateral loads are applied perpendicular to walls or cladding. Bracing these are the roof and floor diaphragms, which transfer their loads to the walls parallel to the load. Walls are supported by the ground. The weight of the structure (and sometimes deep foundations) keeps the system from tipping over.

Connections are critical to complete load paths. We need to ensure the lateral loads flow from perpendicular walls and floors, into diaphragms, into walls parallel to the load, and down to the foundation. Each time the load enters a new element, there must be a connection. The details in Sections 8.4.5 and 8.5.5 show how this is done.

Connections from light to heavier materials warrant special consideration, particularly roof–wall interfaces. Figure 8.4a shows a common, seismically deficient, connection between a concrete wall and wood roof. As the wall moves away from the sheathing, the 2× is placed in cross-grain bending, resulting in failure. Without much additional effort, we can connect the wall to a metal strap and blocking, as in Figure 8.4b, and get a connection that will keep the wall from tipping over.

8.3 SEISMIC DESIGN CONSIDERATIONS

Building codes limit different seismic systems and their maximum heights to ensure the structures perform well during an earthquake. Table 8.1 presents different timber lateral systems and their permissible heights. It also includes the response modification factor, R, which reduces the seismic design force, as a function of energy absorption. A higher R indicates a better performing seismic system.

8.4 DIAPHRAGMS

Lateral systems include horizontal and vertical elements. Horizontal systems consist of diaphragms and **drag struts** (**collectors**). Vertical elements consist of shear walls and frames. Horizontal systems transfer forces through connections to vertical elements, which carry the loads into the foundation.

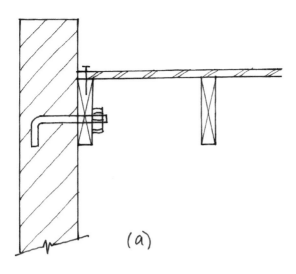

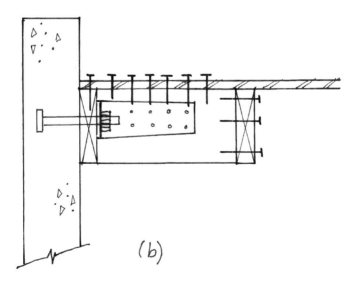

Figure 8.4 (a) Ineffective and (b) effective concrete wall to wood roof connection

Timber Lateral Design   173

Table 8.1 Seismic lateral system *R* factors and maximum heights

Seismic Force Resisting System	Response Coefficient	Permitted Height for Seismic Category					
	R	B	C	D	E	F
Timber							
Light frame walls, structural panel sheathed walls	6½	NL	NL	65	65	65
Light frame walls, other panel sheathed walls	2	NL	NL	35	NP	NP
Light frame walls with flat strap bracing	4	NL	NL	65	65	65
Concrete							
Special moment frames	8	NL	NL	NL	NL	NL
Special reinforced shear walls	5	NL	NL	160	160	100
Steel							
Special moment frames	8	NL	NL	NL	NL	NL
Special concentrially braced frames	6	NL	NL	160	160	100
Masonry							
Special reinforced shear walls	5	NL	NL	160	160	100

Notes: NL = no limit; NP = not permitted

Source: ASCE 7-10

Paul W. McMullin

Diaphragms consist of structural panels and straight or diagonal sheathing boards. In other structures, they are made of concrete slabs, bare metal deck, and diagonal bracing. Timber diaphragms have comparatively low capacity. However, for smaller structures, or those with light walls, timber diaphragms perform well. Diaphragms make possible large, open spaces, without internal walls—so long as there is adequate vertical support.

8.4.1 Forces

We can visualize diaphragms as deep beams that resist lateral loads, as illustrated in Figure 8.5a. They experience maximum bending forces near their middle and maximum shear at their supports (where they connect to walls or frames), as seen in Figure 8.5b.

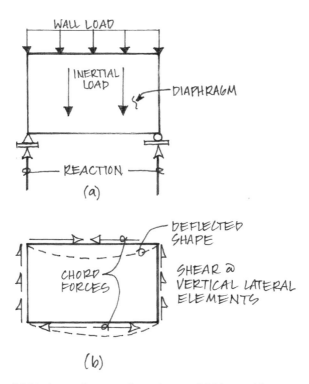

Figure 8.5 (a) Diaphragm forces and reactions, and (b) internal forces

We resolve the mid-span moments into a tension–compression **couple**, requiring boundary elements around their edges. These may be beams, joists, or wall top plates. Often, a double wall top plate (with staggered splices) can resist these forces, as the distance between them is large.

Shear forces are distributed throughout the length of the diaphragm in the direction of lateral force. Because many shear walls and frames do not go the length of the building, the transfer of shear forces between the diaphragm and vertical elements causes high stress concentrations at the ends of the wall or frame, as illustrated in Figure 8.6a. By adding drag struts (also called collectors), we gather the shear stresses into this stronger element, which can then deliver the force to the wall or frame. This reduces the stress concentration (Figure 8.6b) and ensures the

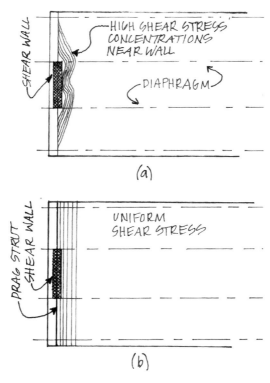

Figure 8.6 Diaphragm stress distribution, (a) without, and (b) with, drag struts

Paul W. McMullin

diaphragm retains its **integrity**. Drag struts frequently consist of beams, joists, and metal straps in timber structures. Note that a structural element that acts as a drag strut will act as a diaphragm chord when the forces are turned 90° and analyzed.

8.4.2 Geometric Considerations

To ensure reasonable behavior, the *NDS*[1] limits diaphragm **aspect ratios** (L/W) to those in Table 8.2. We can use this table when laying out shear walls, to ensure the diaphragms are well proportioned.

There are times when cantilevering a portion of the diaphragm is advantageous, to provide larger open spaces. For instance, occupancies with high concentrations of openings on one side—stores and hotels—typically have large shear walls on the remaining three sides, or down the middle, as illustrated in Figure 8.7. However, cantilevered diaphragms come with the following limitations:

- Diaphragms must be sheathed with structural panels or diagonal lumber.
- Aspect ratios (L'/W') are limited to those in Table 8.3.

Table 8.2 Diaphragm aspect ratio limits

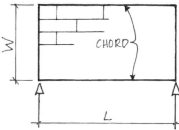

Sheathing Type	Max L/W Ratio
Structural panel, unblocked	3:1
Structural panel, blocked	4:1
Single straight lumber sheathing	2:1
Single diagonal lumber sheathing	3:1
Double diagonal lumber sheathing	4:1

Source: NDS 2015

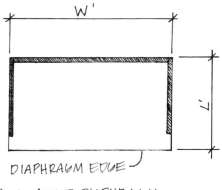

(a) 3-SIDED DIAPHRAGM

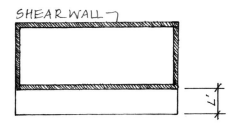

(b) CANTILEVERED DIAPHRAGM

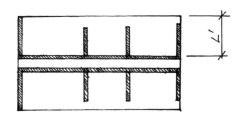

(c) INTERIOR SHEAR WALLS

Figure 8.7 Cantilevered diaphragm types

Paul W. McMullin

Table 8.3 Cantilevered diaphragm aspect ratio limits

Sheathing Type	Max L'/W' Ratio
Wind or seismic torsionally regular:	
Structural panels	1.5:1
Diagonal sheathing	1:1
Open front, seismic torsionally irregular:	
Single story	1:1
2 or more stories	0.67:1

Source: NDS 2015

- Diaphragms must be considered rigid or semi-rigid.
- Diaphragms must meet the *ASCE 7*[2] **drift** requirements for seismic loads.
- L' must be less than 35 ft (10.7 m).

8.4.3 Analysis

To design a timber diaphragm, we need to know the shear and moment distribution in it—though often just the maximum shear and moment. We take these and find the unit shear and tension–compression couple. The steps are as follows, illustrated in Figure 8.8:

- Draw the diaphragm and dimensions *L* and *W*.
- Apply forces from the walls and floor as a line load, *w*.
- Draw the shear and moment diagrams. For a simply supported diaphragm, the maximum shear and moment are:

$$V = \frac{wL^2}{2}$$
(8.1)

$$M = \frac{wL^2}{8}$$
(8.2)

where:

w = uniformly distributed load from walls and floors, lb/ft (kN/m)

L = span, ft (m)

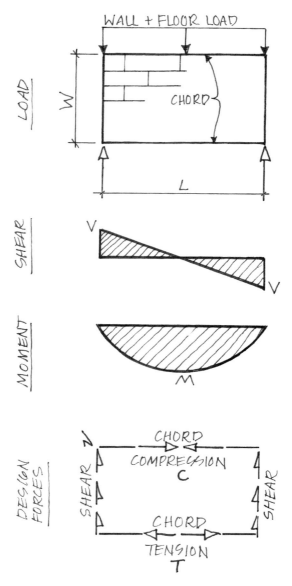

Figure 8.8 Diaphragm shear and moment diagram and design forces

Paul W. McMullin

- Calculate the unit shear, v, by dividing the shear force, V, by the depth, W:

$$v = \frac{V}{W}$$

(8.3)

- Convert the moment, M, to a tension–compression couple, as follows:

$$T = \frac{M}{W} \qquad C = \frac{M}{W}$$

(8.4)

For the perpendicular direction, we follow the previous steps, rotating the load and dimension labels by 90°.

8.4.4 Capacity

Knowing the diaphragm forces, we can choose the sheathing and size the chords.

Diaphragm shear strength is a function of sheathing orientation and nailing patterns. Table 8.4 presents allowable diaphragm strengths for various sheathing types and configurations in pounds per foot (kN/m). Enter the table and find a value higher than the demand, v. The column heading will tell you the nail pattern, and the information to the left will tell you sheathing thickness and nail size. As long as the allowable strength is greater than the demand, v, you are good to go.

To size the chords, we take the tension and compression demands and follow the provisions of Chapters 3 and 6.

8.4.5 Detailing

Structural performance, particularly in earthquakes, depends on detailing. Timber is vulnerable to poor detailing, because we take many small pieces and connect them together. Figures 8.9–8.12 show common diaphragm details.

The following things should be kept in mind:

- Edge nailing is between panel edges, not the diaphragm edge.
- Boundary nailing occurs at the interface with shear walls or frames—which define the diaphragm boundary. You may have a diaphragm boundary in the middle of a structure.
- Blocking connects adjacent panel edges together and greatly enhances shear strength.

Table 8.4 Allowable seismic diaphragm shear strength

Imperial Units

Sheathing Grades	Nail Size	Panel Thickness (in)	Unblocked Case 1 6 in Spacing (lb/ft)	Case 2 6 in Spacing (lb/ft)	Blocked Nail Spacing 1 (in) 6 / Nail Spacing 2 (in) 6 (lb/ft)	4 / 4 (lb/ft)	2½ / 2½ (lb/ft)	2 / 2 (lb/ft)
Structural I	6d	$^5/_{16}$	165	125	185	250	375	420
	8d	$^3/_8$	240	180	270	360	530	600
	10d	$^{15}/_{32}$	285	215	320	425	640	730
Sheathing, single floor	6d	$^5/_{16}$	150	110	170	225	335	380
		$^3/_8$	165	125	185	250	375	420
	8d	$^3/_8$	215	160	240	320	480	545
		$^7/_{16}$	230	170	255	340	505	575
		$^{15}/_{32}$	240	180	270	360	530	600
	10d	$^{15}/_{32}$	255	190	290	385	575	655
		$^{19}/_{32}$	285	215	320	425	640	730

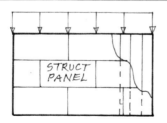 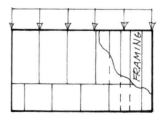

CASE 1 SHEATHING
CASES 2-6 ANY OTHER CONFIGURATION

Notes: (1.) Do not adjust with C_D; (2.) 2 in nominal framing member width; (3.) Spacing 1—at boundaries, panel edges parallel to load in Cases 3 & 4, all panel edges, Cases 5 & 6; (4.) Spacing 2—at other panel edges; (5.) for wind only, multiply values by 1.4

Paul W. McMullin

Table 8.4 cont.

Metric Units

Sheathing Grades	Nail Size	Panel Thickness (mm)	Unblocked Case 1 150 mm Spacing (kN/m)	Case 2 150 mm Spacing (kN/m)	Blocked Nail Spacing 1 (mm) 150 Nail Spacing 2 (in) 150 (kN/m)	100 150 (kN/m)	65 100 (kN/m)	50 75 (kN/m)
Structural I	6d	7.9	2.41	1.82	2.70	3.65	5.47	6.13
	8d	9.5	3.50	2.63	3.94	5.25	7.73	8.76
	10d	11.9	4.16	3.14	4.67	6.20	9.34	10.7
Sheathing, single floor	6d	7.9	2.19	1.61	2.48	3.28	4.89	5.55
		9.5	2.41	1.82	2.70	3.65	5.47	6.13
	8d	9.5	3.14	2.34	3.50	4.67	7.01	7.95
		11.1	3.36	2.48	3.72	4.96	7.37	8.39
		11.9	3.50	2.63	3.94	5.25	7.73	8.76
	10d	11.9	3.72	2.77	4.23	5.62	8.39	9.56
		15.1	4.16	3.14	4.67	6.20	9.34	10.7

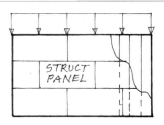
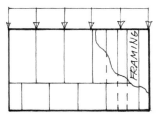

CASE 1 SHEATHING
CASES 2-6 ANY OTHER CONFIGURATION

Notes: (1.) Do not adjust with C_D; (2.) 50 mm nominal framing member width; (3.) Spacing 1—at boundaries, panel edges parallel to load in Cases 3 & 4, all panel edges, Cases 5 & 6; (4.) Spacing 2—at other panel edges; (5.) for wind only, multiply values by 1.4

Source: NDS 2015

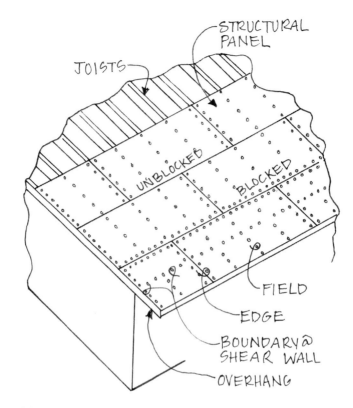

Figure 8.9 Diaphragm nailing pattern

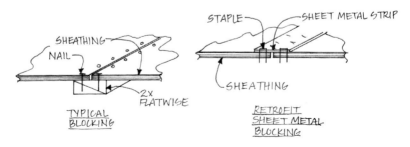

Figure 8.10 Diaphragm blocking

Paul W. McMullin

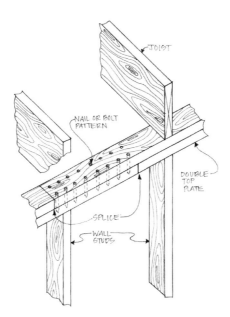

Figure 8.11 Top plate chord splice

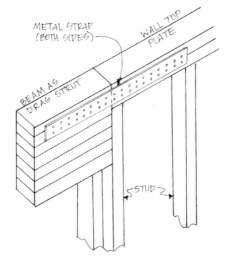

Figure 8.12 Drag strut to wall connection

Timber Lateral Design 185

8.5 SHEAR WALLS

Shear walls—with structural sheathing—are the commonest vertically oriented lateral systems in timber structures. In older structures, we often straight board sheathing or diagonal bracing.

8.5.4 Forces

Shear walls resist lateral loads through horizontal shear in the sheathing and tension–compression couples in the end studs, as shown in Figure 8.13. They are efficient and stiff and particularly suited to buildings with partitions and perimeter walls and without an overabundance of windows. Short, tall shear walls concentrate forces at their base, requiring large foundations, whereas long, short walls reduce footing loads.

8.5.2 Geometric Considerations

Like diaphragms, we limit shear wall aspect ratios, h/b_s, as shown in Table 8.5. For non-seismic applications, they are more permissive than diaphragms. However, if the structure is in a seismic area, the aspect ratio is limited to 2, without strength being reduced. (This will inform the height of windows in walls that resist lateral shear.)

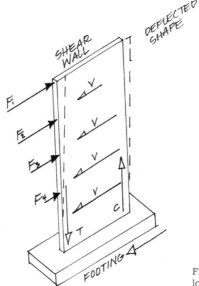

Figure 8.13 Shear wall external lateral loads and internal forces

Paul W. McMullin

Table 8.5 Shear wall aspect ratio limits

Sheathing Type	Max h/b_s Ratio
Structural panel, blocked, non-seismic	3.5:1
Structural panel, blocked, seismic	2:1
Particleboard, blocked	2:1
Diagonal sheathing	2:1
Gypsum wallboard	2:1
Portland cement plaster	2:1
Fiberboard	1.5:1

Source: NDS 2015

Wall height, h, and width, b_s, are defined in two ways: raw geometry and force transfer, as illustrated in Figure 8.14. Most commonly, height and width are based on raw geometry. In this case, the height is the floor height, and width is the distance between openings, shown in Figure 8.14a.

When we need to push the design requirements, we use the force transfer definitions, illustrated in Figure 8.14b. In this case, the width is the same as for raw geometry. However, the height is the opening height. This allows us to meet the requirements of Table 8.5 with narrower wall segments. The caveat is that we must reinforce around the openings to ensure that the forces transfer properly around the opening, as shown in Figure 8.15.

8.5.3 Analysis

To analyze a shear wall, we need to know how the lateral forces are distributed between wall segments in the same wall. Although we could base it on stiffness, we typically use wall length. For the three-segment wall shown in Figure 8.16, we have three different wall lengths. The force in each length is equal to the applied shear force times the length in question, divided by the sum of the lengths. For the first wall segment, we get:

$$V_1 = \frac{Vb_{s1}}{b_{s1} + b_{s2} + b_{s3}}$$

(8.5)

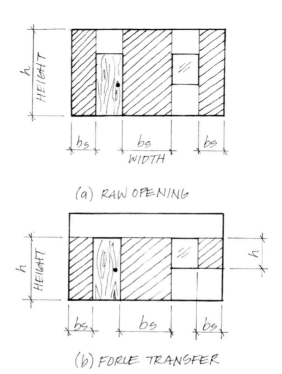

(a) RAW OPENING

(b) FORCE TRANSFER

Figure 8.14 Shear wall geometry definitions

where

    V = total force to wall line (k, kN)

    V_1 = force to wall segment 1 (k, kN)

    b_{s1}, b_{s2}, b_{s3} = wall segment widths (ft, m)

Once we determine the forces in each segment, we can analyze the segments as stand-alone walls, as discussed in the next paragraphs and shown in Figure 8.16b.

In a shear wall, the shear force is constant from top to bottom. The moment is a maximum at the bottom and zero at the top. Figure 8.17 illustrates the applied forces, shear and moment diagrams, and design forces.

            Paul W. McMullin

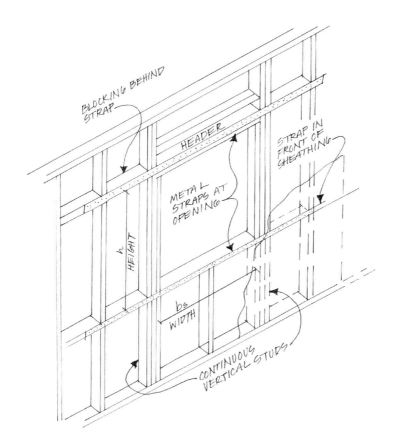

Figure 8.15 Force transfer reinforcing around an opening

The steps to analyze a segment of shear wall are as follows:

- Draw the wall with the applied shear force, reactions, and dimensions h and b_s;
- Calculate the moment at the base of the wall as:

$$M = V_h \tag{8.6}$$

where:

V = shear force at the top of the wall, lb (kN)
h = wall height, ft (m)

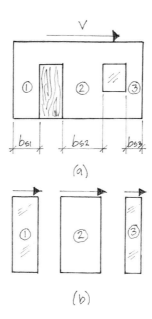

Figure 8.16 Shear wall force distribution

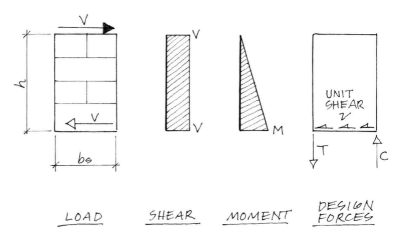

Figure 8.17 Shear wall shear and moment diagrams and design forces

Paul W. McMullin

- Calculate the unit shear by dividing the shear force, V, by the width, b_s:

$$v = \frac{V}{b_s}$$
(8.7)

- Convert the moment M to a tension–compression couple, as follows:

$$T = \frac{M}{0.9b_s} \qquad\qquad C = \frac{M}{0.9b_s}$$
(8.8)

The 0.9 accounts for the fact that the centroid of the compression studs and bolt in the hold down are not at the ends of the wall. We can calculate a more accurate value once we draw the wall geometry (shown in the example).

8.5.4 Capacity

Knowing the shear wall forces, we can choose the sheathing and size the end studs.

Shear wall strength is a function of sheathing orientation and nailing patterns. Table 8.6 provides allowable shear wall strengths for various sheathing types and configurations in pounds per foot (kN/m) of width. Enter the table and find the wall strength you need. The column heading will tell you the nail spacing, and the information to the left indicates sheathing thickness and nail size. Remember that all sheathing joints must be blocked in a shear wall. As long as the allowable strength is greater than the demand, v, the sheathing has adequate strength.

To size the end studs, we take the tension and compression demands and follow the provisions of Chapters 3 and 6. It is important to check the cross-grain compression where the stud sits on a sill plate or other framing.

8.5.5 Detailing

Just like diaphragms, shear walls must be well detailed to ensure adequate performance. The following details show how we detail walls:

- sheathing nailing—see Figure 8.18;
- blocking—see Figure 8.10;
- hold down between floors—see Figure 8.19;
- hold down to concrete—Figure 8.20.

Table 8.6 Allowable seismic shear wall sheathing shear strength

Imperial Units

Sheathing Grade	Nail Size	Panel Thickness (in)	Nail Spacing (in)			
			6 (lb/ft)	4 (lb/ft)	3 (lb/ft)	2 (lb/ft)
Structural I	6d	$5/16$	200	300	390	510
	8d	$3/8$	230	360	460	610
		$7/16$	255	395	505	670
		$15/32$	280	430	550	730
	10d	$15/32$	340	510	665	870
Sheathing	6d	$5/16$	180	270	350	450
		$3/8$	200	300	390	510
	8d	$3/8$	220	320	410	530
		$7/16$	240	350	450	585
		$15/32$	260	380	490	640
	10d	$15/32$	310	460	600	770
		$19/32$	340	510	665	870
Board Sheathing	8d					
• Horizontal Lumber						50
• Diagonal Lumber						300
• Double Diagonal Lumber						600

Notes: (1.) Do not adjust with C_D; (2.) 2 in nominal framing member width;
(3.) all panel edges must be blocked; (4.) for wind only, multiply values by 1.4;
(5.) lumber sheathing is 1 in nominal thickness

Paul W. McMullin

Table 8.6 cont.

Metric Units

Sheathing Grade	Nail Size	Panel Thickness (mm)	Nail Spacing (mm)			
			150 (kN/m)	100 (kN/m)	75 (kN/m)	50 (kN/m)
Structural I	6d	7.9	2.92	4.38	5.69	7.44
	8d	9.5	3.36	5.25	6.71	8.90
		11.1	3.72	5.76	7.37	9.78
		11.9	4.09	6.28	8.03	10.65
	10d	11.9	4.96	7.44	9.70	12.70
Sheathing	6d	7.9	2.63	3.94	5.11	6.57
		9.5	2.92	4.38	5.69	7.44
	8d	9.5	3.21	4.67	5.98	7.73
		11.1	3.50	5.11	6.57	8.54
		11.9	3.79	5.55	7.15	9.34
	10d	11.9	4.52	6.71	8.76	11.24
		15.1	4.96	7.44	9.70	12.70
Board Sheathing	8d					
• Horizontal Lumber						0.73
• Diagonal Lumber						4.38
• Double Diagonal Lumber						8.76

Notes: (1.) Do not adjust with C_D; (2.) 50 mm nominal framing member width; (3.) all panel edges must be blocked; (4.) For wind only, multiply values by 1.4; (5.) lumber sheathing is 1 in nominal thickness

Source: NDS 2015

Some things to keep in mind:

- All panel edges of shear walls must be blocked.
- We need a load path from the diaphragm to shear wall. This is often done through rim joists or blocking.
- Most timber structures are light enough that they need hold downs in the ends of shear walls to prevent uplift.

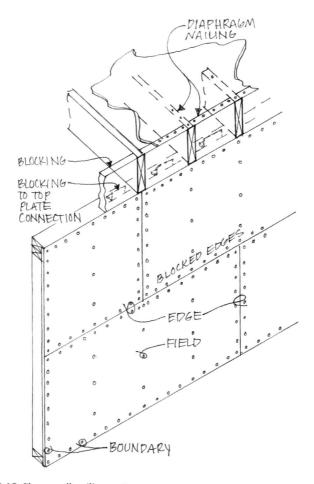

Figure 8.18 Shear wall nailing pattern

Paul W. McMullin

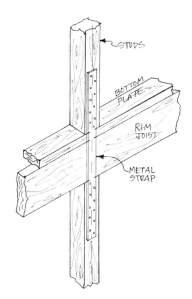

Figure 8.19 Hold down between floors

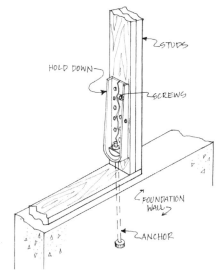

Figure 8.20 Hold down to concrete

Timber Lateral Design   195

8.6 SHEAR WALL EXAMPLE

Step 1: Draw structural layout

We begin by sketching out the area of interest sufficiently so we know what the wall supports. Figure 8.21 shows an elevation view of the wall, with the applied lateral loads. Taking portion A, we draw its free body diagram, applied loads, and reactions, shown in Figure 8.22. Next, we need a cross section of the wall to get information for hold-down calculations, shown in Figure 8.23.

Step 2: Loads

With the structural layout and wall geometry, we are ready to determine the loads for overturning. The governing load combination is $0.6D + 0.7E$, which minimizes dead load. We will use this to calculate shear and downward weight. Note we don't need to add them together.

Step 2a: Lateral Loads

We determine V_R and V_3 from the seismic **base shear** and vertical distribution, which is a function of weight at each story. The analysis is beyond the scope of this example, so we will take them as:

V_R = 10 k	V_R = 44.5 kN
V_3 = 17 k	V_3 = 76 kN

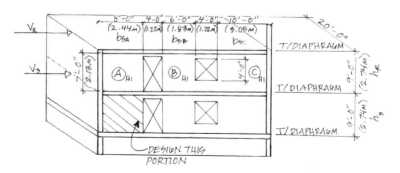

Figure 8.21 Shear wall example layout

Paul W. McMullin

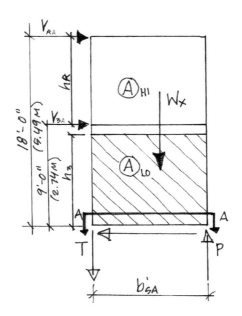

Figure 8.22 Shear wall free body diagram

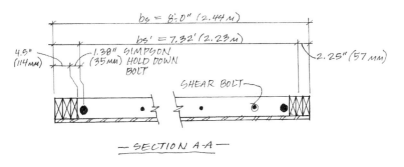

Figure 8.23 Shear wall base section (A-A)

Timber Lateral Design   197

We multiply these by 0.7 to get the ASD level:

$V_R = 10 \text{ k}(0.7)$ $V_R = 44.5 \text{ kN}(0.7)$

    $= 7 \text{ k}$ $= 31.1 \text{ kN}$

$V_3 = 17 \text{ k}(0.7)$ $V_3 = 76 \text{ kN}(0.7)$

    $= 11.9 \text{ k}$ $= 53.2 \text{ kN}$

We next determine the lateral shear in wall segment A. We do this based on length (though stiffness would be more exact). Key lengths are:

$b_{sA} = 8 \text{ ft}$ $b_{sA} = 2.44 \text{ m}$

$b_{sT} = 8 \text{ ft} + 6 \text{ ft} + 10 \text{ ft} = 24 \text{ ft}$ $b_{sT} = 2.44 \text{ m} + 1.83 \text{ m} + 3.05 \text{ m} = 7.3 \text{ m}$

$$V_{RA} = \frac{V_R b_{sA}}{b_{sT}}$$

$$= \frac{(7 \text{ k})(8 \text{ ft})}{24 \text{ ft}} \qquad\qquad = \frac{(31.1 \text{ kN})(2.44 \text{ m})}{7.3 \text{ m}}$$

$$= 2.33 \text{ k} \qquad\qquad\qquad\quad = 10.4 \text{ kN}$$

$$V_{3A} = \frac{V_3 b_{sA}}{b_{sT}}$$

$$= \frac{(11.9 \text{ k})(8 \text{ ft})}{24 \text{ ft}} \qquad\qquad = \frac{(53.2 \text{ kN})(2.44 \text{ m})}{7.3 \text{ m}}$$

$$= 3.97 \text{ k} \qquad\qquad\qquad\quad = 17.8 \text{ kN}$$

To size the sheath, we will need to know the unit shear in the wall. We do this by dividing the wall shear by its length:

$$v_{3A} = \frac{V_{RA} + V_{3A}}{b_{sA}}$$

$$= \frac{2.33 \text{ k} + 3.97 \text{ k}}{8 \text{ ft}} \qquad\qquad = \frac{10.4 \text{ kN} + 17.8 \text{ kN}}{2.44 \text{ m}}$$

$$= 0.788 \frac{\text{k}}{\text{ft}} \qquad\qquad\qquad = 11.6 \frac{\text{kN}}{\text{m}}$$

Step 2b: Gravity Loads

We need the dead load tributary to the wall segment to calculate the stabilizing moment. Because the joists run parallel to the wall, we only consider the weight of the wall itself. Taking $q_D = 10 \text{ lb/ft}^2$, we find:

                Paul W. McMullin

$W = 0.6q_DA_W$ (Note, the 0.6 is from the overturning load combination)

$= 0.6(10 \text{ lb/ft}^2)(8 \text{ ft})(18 \text{ ft})$ $= 0.6(0.48 \text{ kN/m}^2)(2.4 \text{ m})(5.5 \text{ m})$

$= 864 \text{ lb}$ $= 3.8 \text{ kN}$

Step 2c: Moments

We now find the overturning and stabilizing moments.

$M_{ot} = V_{RA}h_{RA} + V_{3A}h_{3A}$

$= (2.23\text{k})(18 \text{ ft}) + (3.97 \text{ k})(9 \text{ ft})$ $= (10.4 \text{ kN})(5.48 \text{ m}) + (17.8 \text{ kN})(2.74 \text{ m})$

$= 77.67\text{k} - \text{ft}$ $= 105.8 \text{ kN} - \text{m}$

$M_{st} = W(b'_{sA}/2)$

$= 0.86 \text{ k}(7.32 \text{ ft}/2)$ $= 3.8 \text{ kN}(2.23 \text{ m}/2)$

$= 3.15 \text{ k} - \text{ft}$ $= 4.24 \text{ kN} - \text{m}$

Now, to find reactions at wall ends:

$T = \dfrac{M_{ot} - M_{st}}{b'_{sa}}$

$= \dfrac{77.67 \text{ k} - \text{ft} - 3.15 \text{ k} - \text{ft}}{7.32 \text{ ft}}$ $= \dfrac{105.8 \text{ kN} - \text{m} - 4.24 \text{ kN} - \text{m}}{2.23 \text{ m}}$

$= 10.2 \text{ k}$ $= 45.5 \text{ kN}$

$P = \dfrac{M_{ot} + M_{st}}{b'_{sa}}$

$= \dfrac{77.67 \text{ k} - \text{ft} + 3.15 \text{ k} - \text{ft}}{7.32 \text{ ft}}$ $= \dfrac{105.8 \text{ kN} - \text{m} + 4.24 \text{ kN} - \text{m}}{2.23 \text{ m}}$

$= 11.0 \text{ k}$ $= 49.3 \text{ kN}$

This now gives us the forces we need to design the sheathing and member at the end of the walls.

Step 3: Material Parameters

We will use wood structure panels and determine their thickness later.

2× material is Douglas Fir–Larch (N) stud grade with the following properties:

$$F_t = 400\,\frac{\text{lb}}{\text{in}^2}\left(2{,}758\,\frac{\text{kN}}{\text{m}^2}\right)$$

from Table A2.1 in Appendix 2

$F_c = 900\ \text{lb/in}^2\ (6{,}205\ \text{kN/m}^2)$

$F_{c\perp} = 625\ \text{lb/in}^2\ (4{,}309\ \text{kN/m}^2)$

$$E_{\min} = 510\,\frac{\text{k}}{\text{in}^2}\left(3{,}516\,\frac{\text{MN}}{\text{m}^2}\right)$$

We get the adjusted reference values from:

$F'_t = F_t C_D C_M C_t C_F C_i$

$F'_c = F_c C_D C_M C_t C_F C_i C_p$

$F'_{c\perp} = F_{c\perp} C_M C_t C_i C_b$

$E'_{\min} = E_{\min} C_M C_t C_i C_T$

The easy adjustment factors are as shown in the table.

Factor	Description	Source
$C_D = 1.6$	Load duration	Table A4.1
$C_M = 1.0$	Wet service	Table A4.2
$C_t = 1.0$	Temperature	Table A4.4
$C_F = 1.3$	Size—tension	Table A4.5
$C_F = 1.1$	Size—compression	
$C_i = 1.0$	Incising no treatment	Table A4.10
$C_b = 1.0$	Bearing, close to end	Table A4.11
$C_T = 1.0$	Buckling stiffness—not a truss chord	Section 2.4.13

We will save C_p for later. Finding the adjusted design values that we can at this point:

$$F'_t = 400\,\frac{\text{lb}}{\text{in}^2}(1.6)\,(1.0)\,(1.3)$$
$$= 832\,\frac{\text{lb}}{\text{in}^2}$$

$$F'_t = 2{,}758\,\frac{\text{kN}}{\text{m}^2}(1.6)\,(1.0)\,(1.3)$$
$$= 5{,}737\,\frac{\text{kN}}{\text{m}^2}$$

$$F'_{c\perp} = 625\,\frac{\text{lb}}{\text{in}^2}(1.0) = 625\,\frac{\text{lb}}{\text{in}^2}$$

$$F'_{c\perp} = 4{,}309\,\frac{\text{kN}}{\text{m}^2}(1.0) = 4{,}309\,\frac{\text{kN}}{\text{m}^2}$$

Paul W. McMullin

$$E'_{min} = 510\frac{k}{in^2}(1.0) = 510\frac{k}{in^2} \qquad E'_{min} = 3{,}516\frac{MN}{m^2}(1.0) = 3{,}516\frac{MN}{m^2}$$

It will also be helpful to find the adjusted design compressive stress, without C_P:

$$F^*_c = F_c C_D C_M C_t C_F C_i$$

$$F^*_c = 900\frac{lb}{in^2}(1.6)(1.0)(1.1) \qquad F^*_c = 6{,}205\frac{kN}{m^2}(1.6)(1.0)(1.1)$$

$$= 1{,}584\frac{lb}{in^2} \qquad\qquad\qquad = 10{,}920\frac{kN}{m^2}$$

Step 4: Initial Sizes

We will find the sheath thickness directly from strength—Table 8.6.

For studs at the end, we will base our initial area on compression perpendicular to the grain:

$$A_{req} = \frac{P}{0.5\,(F^*_c)}$$

$$= \frac{11.0\,k}{0.5(1{,}584\,lb/in^2)}\frac{1000\,lb}{1\,k} \qquad = \frac{49.3\,kN}{0.5(10{,}920\,kN/m^2)\left(1\,m/_{1000\,mm}\right)^2}$$

$$= 13.9\,in^2 \qquad\qquad\qquad = 9{,}029\,mm^2$$

Before choosing the number of end studs, we check cross grain bearing. Taking $F'_{c\perp}$ from before, we find the required bearing area:

$$A_{req} = \frac{P}{F'_{c\perp}}$$

$$= \frac{11.0\,k}{625\,lb/in^2}\left(\frac{1000\,lb}{1\,k}\right) \qquad = \frac{49.3\,kN}{(4{,}309\,kN/m^2)\left(1\,m/_{1000\,mm}\right)^2}$$

$$= 17.6\,in^2 \qquad\qquad\qquad = 11{,}440\,mm^2$$

We will use a 2×6 (50×150) stud to allow for increased insulation. Dividing area by the stud depth, 5.5 in (140 mm), we get the total width:

$$b = \frac{A_{req}}{d}$$

$$= \frac{17.6\,in^2}{5.5\,in} = 3.2\,in \qquad\qquad = \frac{11{,}440\,mm^2}{140\,mm} = 82\,mm$$

This equals three studs.

Step 5: Strength

Step 5a: Sheathing

Knowing the shear demand, we can find the sheath thickness and nailing required directly from Table 8.6, which already has the $C_D = 1.6$ factor included. Entering the table, we look for a value slightly larger than the shear demand, v_{3A}. However, in this case, there isn't a value high enough without going to a very tight nail spacing. We will have to go to a wall with sheathing on both sides.

This means our shear demand is half v_{3A} for each side, 394 lb/ft (5.8 kN/m). Entering Table 8.6 again, we see several options. In this case, let's select the value of $v_a = 450$ lb/ft (6.57 kN/m), in the fourth row from the bottom. This corresponds to an 8d nail spacing of 3 in (75 mm) and $\frac{7}{16}$ in (11 mm) siding thickness—which is economical.

Step 5b: Sill Plate Compression

To find compression stress, we need the gross compression area of the studs:

$$A_g = bd$$
$$= 3(1.5 \text{ in}) 5.5 \text{ in} \qquad\qquad = 3(38 \text{ mm}) 140 \text{ mm}$$
$$= 24.75 \text{ in}^2 \qquad\qquad\qquad = 15,960 \text{ mm}^2$$

Calculating bearing stress, we get:

$$f_{c\perp} = \frac{P}{A_G}$$

$$= \frac{11.0 \text{ k}}{24.75 \text{ in}^2}\left(\frac{1000 \text{ lb}}{1 \text{ k}}\right) \qquad\qquad = \frac{49.3 \text{ kN}}{0.016 \text{ m}^2}$$

$$= 444 \text{ lb/in}^2 < F_{c\perp} \qquad\qquad = 3{,}081 \text{ kN/m}^2 < F_{c\perp}$$

Again, this confirms we are OK.

Step 5c: Stud Compression

We now check the studs in compression. Our initial stud quantity was based on compression perpendicular to grain, and so we don't need to check this again. But, we do need to look at stress parallel to the grain.

We begin by finding the column stability factor, C_p.

Paul W. McMullin

$$C_P = \frac{1 + \left(F_{CE}/F_C^*\right)}{2c} - \sqrt{\left[\frac{1 + \left(F_{CE}/F_C^*\right)}{2c}\right]^2 - \frac{\left(F_{CE}/F_C^*\right)}{c}}$$

We have a good portion of what we need from above, but will fill in a few more terms before taking on the equation.

$$F_{CE} = \frac{0.822 E'_{min}}{\left(l_e/d\right)^2}$$

l_e = 8 ft (2.44 m)

We check column slenderness:

$$\frac{l_e}{d} = \frac{8 \text{ ft}}{5.5 \text{ in}}\left(\frac{12 \text{ in}}{1 \text{ ft}}\right) = 17.5 \qquad \frac{l_e}{d} = \frac{2.44 \text{ m}}{140 \text{ mm}}\left(\frac{1000 \text{ mm}}{1 \text{ m}}\right) = 17.5$$

$$F_{CE} = \frac{0.822\left(510 \text{ k/in}^2\right)}{(17.5)^2}\left(\frac{1000 \text{ lb}}{1 \text{ k}}\right) = 1{,}369 \frac{\text{lb}}{\text{in}^2}$$

$$F_{CE} = \frac{0.822\left(3{,}516 \text{ MN/m}^2\right)}{(17.5)^2}\left(\frac{1000 \text{ kN}}{1 \text{ MN}}\right) = 9{,}437 \frac{\text{kN}}{\text{m}^2}$$

c = 0.8 for sawn lumber

Now, putting it all together, with an intermediate step to help keep things sane:

$$C_P = \frac{1 + \left(1{,}369/1{,}584\right)}{2(0.8)} - \sqrt{\left[\frac{1 + \left(1{,}369/1{,}584\right)}{2(0.8)}\right]^2 - \frac{\left(1{,}369/1{,}584\right)}{(0.8)}}$$

$$= 1.165 - \sqrt{(1.165)^2 - 1.080}$$

$$= 0.64$$

$$C_P = \frac{1+\left(9{,}437\big/10{,}920\right)}{2(0.8)} - \sqrt{\left[\frac{1+\left(9{,}437\big/10{,}920\right)}{2(0.8)}\right]^2 - \frac{\left(9{,}437\big/10{,}920\right)}{(0.8)}}$$

$$= 1.165 - \sqrt{(1.165)^2 - 1.080}$$

$$= 0.64$$

Now, we can calculate the adjusted design compression stress, F_c':

$$F_c' = F_c^* C_P$$

$$= 1{,}584\,\frac{\text{lb}}{\text{in}^2}(0.64) = 1{,}014\,\frac{\text{lb}}{\text{in}^2} \qquad = 10{,}920\,\frac{\text{kN}}{\text{m}^2}(0.64) = 6{,}989\,\frac{\text{kN}}{\text{m}^2}$$

$$f_c = \frac{P}{A_g}$$

$$= \frac{11.0\ \text{k}}{24.75\ \text{in}^2}\,\frac{1000\ \text{lb}}{1\ \text{k}} \qquad\qquad = \frac{49.3\ \text{kN}}{0.01596\ \text{m}^2}$$

$$= 444\,\frac{\text{lb}}{\text{in}^2} \qquad\qquad\qquad = 3{,}089\,\frac{\text{kN}}{\text{m}^2}$$

We see the axial stress, f_c, is less than the adjusted design stress, and so we are OK. We see, in the end, the bearing perpendicular to grain controlled the design.

Step 5d: Stud Tension

Now, we check the end studs in tension. We first need to know the net area of the studs due to the hold down bolts. Assuming a ¾ in (20 mm) bolt, we find:

$$A_n = A_g - b(hole)$$

$$= 24.75\ \text{in}^2 - 3(1.5\ \text{in})0.875\ \text{in} \qquad = 15{,}960\ \text{mm}^2 - 3(38\ \text{mm})22\ \text{mm}$$

$$= 20.8\ \text{in}^2 \qquad\qquad\qquad\qquad = 13{,}450\ \text{mm}^2$$

$$f_t = \frac{T}{A_n}$$

$$= \frac{10.2\ \text{k}}{20.8\ \text{in}^2}\,\frac{1000\ \text{lb}}{1\ \text{k}} = 490\,\frac{\text{lb}}{\text{in}^2} \qquad = \frac{45.5\ \text{kN}}{0.01345\ \text{m}^2} = 3{,}382\,\frac{\text{kN}}{\text{m}^2}$$

Comparing this with F_t', we see the demand, f_t, is less than the strength, and we are OK.

Whew!

Paul W. McMullin

Step 6: Deflection

We will forego the deflection calculation, given its complexity and lack of necessity for this structure.

Step 7: Summary

We use three 2 × 6 (50 × 150 mm) Douglas Fir–Larch (North) stud boundary elements and double sheath $\frac{7}{16}$ in (11 mm) OSB, 10d, at 3 in (75 mm) on center, 2× blocked framing.

If we were to continue, we would design the hold down, tension bolt in the concrete, and shear bolts. But we'll save that for another day.

8.7 WHERE WE GO FROM HERE

This chapter has introduced the general concepts of lateral design. From here, we estimate lateral forces on a structure and, through structural analysis, determine their distribution into diaphragms and shear walls. This yields internal forces, from which we proportion sheathing thickness, and chord sizes. We then detail the structure, paying particular attention to the seismic requirements discussed above.

Seismic lateral design has become increasingly sophisticated in the past two decades. Prescriptive code requirements are giving way to performance-based design (PBD). This allows the owner and designer to pair the earthquake magnitude and structural performance that is consistent with the function of the building. Additionally, engineers are using PBD for more traditional, code-based buildings to reduce material consumption, as discussed in the *Special Topics* volume of this series.

NOTES

1. ANSI/AWC. *National Design Specification (NDS) for Wood Construction* (Leesburg, VA: AWC, 2015).

2. ASCE. *Minimum Design Loads for Buildings and Other Structures.* ASCE/SEI 7-10 (Reston, VA: American Society of Civil Engineers, 2010).

Timber Connections

Chapter 9

Paul W. McMullin

Timber connectors hold it all together. Sheathing to joists, joists to beams, beams to columns, columns to footings, and a myriad of other connections keep buildings together. Early connections utilized overlapping and finger-type joints, such as mortise and tenon (Figure 9.1), half lap, scarf, and other joints. At the close of the 1800s, steel became affordable, and this ushered in the use of nails, bolts, screws, and, recently, engineered metal plates. These have all simplified and sped up wood construction (Figure 9.2)—though they are not as elegant as early joinery.

9.1 CONNECTOR TYPES

In this chapter, we focus on dowel-type connectors (nails, bolts, and lag screws) and engineered metal plate (Simpson™) connectors. We will briefly discuss truss plates, timber **rivets**, split rings, and **shear plate** connectors. Small-diameter wood screws and finishing nails are not considered structural connectors and are not included.

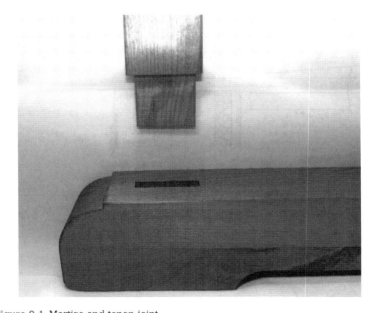

Figure 9.1 Mortise and tenon joint
Source: Photo courtesy of Jason Rapich

Figure 9.2 Nails, lag screws, bolts, and metal plate connectors (left to right, top to bottom)

Paul W. McMullin

9.1.1 Nails and Spikes

Nails are the commonest timber connector and easiest to install. We place two pieces of material together, align the nail, and pound it in with a hammer or nail gun, shown in cut-away form in Figure 9.3. Nails, in the past, were forged by hand or cut from a flat plate billet and were a rarity. Today, they are made from drawn wire and are truly a commodity. Machines pull the wire to the desired diameter and then cut it, forming the tip. The head is formed by a die under pressure. Figure 9.4 shows a variety of nails and a **spike**.

The most-used nail types in construction are **common**, box, and **sinker**, shown in Figure 9.5. Nail guns take nails in a wide range of shapes and sizes. Ring shank and helical nails—Figure 9.6—have deformations on the shaft that give them greater pull-out (withdrawal) resistance. Nail finishes include bright (as-manufactured), coated, and galvanized. The coated and galvanized finishes help keep the nails from pulling out over time or under load. The galvanized finish also increases the resistance to corrosion in wet conditions.

Nail heads vary in style, as shown in Figure 9.7, depending on use. Many have flat, round heads. Sinkers have a taper on the head, allowing them to sit flush to the wood. Heads for nail guns are often clipped or offset so that they fit in the gun, and their diameters vary slightly from their hand-driven counterparts. Nails used for temporary construction, such as concrete formwork, have two heads (**duplex** nails). This allows the nail to be driven to the first head, and then pulled out later with the second head.

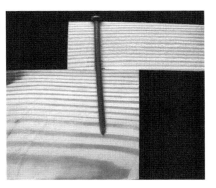

Figure 9.3 Cut-away view of (a) regular and (b) toe-nailed connection

Figure 9.4 Variety of nail types and spike

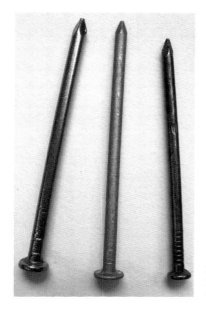

Figure 9.5 Common bright, box galvanized, sinker coated, common nails (from left to right)

Paul W. McMullin

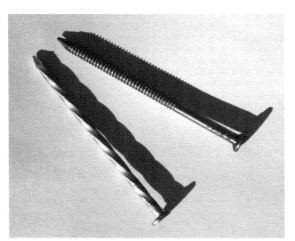

Figure 9.6 Ring shank and helical thread nail

Figure 9.7 Common, box, sinker, spike, duplex, and gun nail head types (left to right)

Table 9.1 Nail sizes for different types

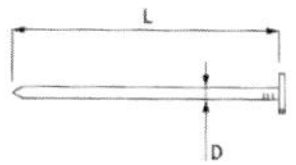

Imperial Units	Nail Dimensions (in)									
		8d	*10d*	*12d*	*16d*	*20d*	*30d*	*40d*	*50d*	*60d*
Common	*L*	2 1/2	3	3 1/4	3 1/2	4	4 1/2	5	5 1/2	6
	D	0.131	0.148	0.148	0.162	0.192	0.207	0.225	0.244	0.263
Box	*L*	2 1/2	3	3 1/4	3 1/2	4	4 1/2	5		
	D	0.113	0.128	0.128	0.135	0.148	0.148	0.162		
Sinker	*L*	2 3/8	2 7/8	3 1/8	3 1/4	3 3/4	4 1/4	4 3/4		5 3/4
	D	0.113	0.120	0.135	0.148	0.177	0.192	0.207		0.244
Metric Units	*Equivalent Nail Dimensions (mm)*									
		8d	*10d*	*12d*	*16d*	*20d*	*30d*	*40d*	*50d*	*60d*
Common	*L*	63.5	76.2	82.6	88.9	101.6	114.3	127.0	139.7	152.4
	D	3.33	3.76	3.76	4.11	4.88	5.26	5.72	6.20	6.68
Box	*L*	63.5	76.2	82.6	88.9	101.6	114.3	127.0		
	D	2.87	3.25	3.25	3.43	3.76	3.76	4.11		
Sinker	*L*	60.3	73.0	79.4	82.6	95.3	108.0	120.7		146.1
	D	2.87	3.05	3.43	3.76	4.50	4.88	5.26		6.20

Nail and spike sizes vary between types. For example, Figure 9.5 shows 8d common, box, and sinker nails. Note the difference in length and diameter. To aid with keeping track of the differences, Table 9.1 gives nail diameters and lengths for common, box, and sinker nails and spikes of varying sizes.

Nail yield strengths range from 80 to 100 k/in^2 (from 550 to 690 MN/m^2). Shear load is transferred by bearing between wood and nail, becomes shear in the nail at the joint interface, and then goes back to bearing in the next piece of wood, as illustrated in Figure 9.8a. Withdrawal load is transferred from the wood to the shaft through friction, and then through the head into the connected member (see Figure 9.8b).

9.1.2 Bolts

We use bolts to join larger members and when loads are higher than a nail group can carry. Bolts consist of a head, a shaft with smooth and threaded portions, washers, and nuts, as seen in Figure 9.9. We install bolts by drilling an equal-diameter hole through the members to be connected and inserting and tightening the bolt, as shown in cut-away form in Figure 9.10. Washers at nuts and hex heads spread the bolt tension out, help to prevent wood damage, and keep the head or nut from pulling into the wood.

Bolt material for timber construction is typically A307 with a 60-k/in^2 (415 kN/m^2) tensile strength. Stronger bolts are of little value, as wood strength generally controls the connector strength. Like nails, force is transferred in bearing between the wood and bolt and through shear in the bolt across the joint, as in Figure 9.8a.

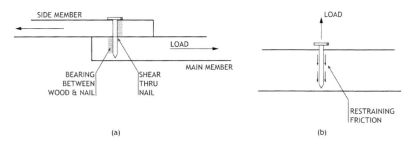

Figure 9.8 Force transfer in nail showing (a) lateral and (b) withdrawal

Figure 9.9 Bolts showing hex and carriage heads (rounded) and zinc-plated and galvanized finishes

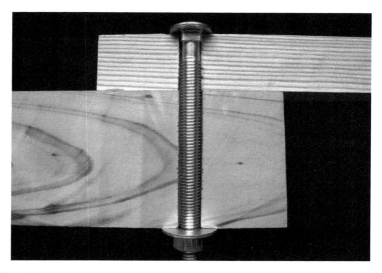

Figure 9.10 Cut-away view of carriage head bolt in single shear connection

Paul W. McMullin

9.1.3 Lag screws

Lag screws join higher load members where a through bolt is impractical. This is often when one cannot access both sides of the member, or the member is so thick a through bolt is unnecessary. Lag screws consist of a hex head, smooth and threaded shaft, as shown in Figure 9.11. Installation begins with a hole slightly smaller than the screw diameter being predrilled for the shaft and threads, as shown in cut-away in Figure 9.12. The threads are coarse 'wood-type' and pull the lag screw into the predrilled hole. Again, a washer against the head protects the wood.

Lag screws carry both lateral and withdrawal loads. Force transfer for lateral loads is the same as in bolts. For withdrawal, load is transferred from the wood into the threads in bearing and shaft in friction into the screw body, then through the head into the connected member.

9.1.4 Engineered Metal Plate Connectors

Engineered connector manufacturers expand scope and applicability every year. Simpson Strong-Tie™ is the best-known manufacturer, though there are others. Connectors include joist hangers, hurricane ties, and angles, shown in Figure 9.13. Column bases and caps (Figure 9.14), heavy

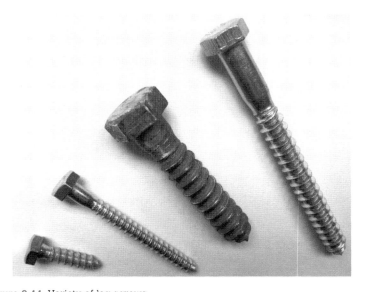

Figure 9.11 Variety of lag screws

Figure 9.12 Cut-away view of a lag screw connection with steel side plate

Figure 9.13 Joist hanger, hurricane tie, and A35 Simpson™ engineered metal plate connectors (from left to right)

Paul W. McMullin

beam connectors, and shear wall hold downs (Figure 9.15), are popular. Each year, manufacturers make new connectors that fit more and more applications. If you work with timber structures, a yearly review will keep you abreast of the new products and how they might benefit your projects.

Figure 9.14 Engineered metal column base connection

Figure 9.15 Shear wall hold down

Figure 9.16 Truss plate (a) showing teeth, and (b) installed on a light framed truss

Source: Image courtesy of Simpson Strong-Tie ©

Paul W. McMullin

9.1.5 Truss Plates

Light wood trusses today are fabricated using truss plates (formerly called connector plates), shown in Figure 9.16. They consist of galvanized, gage metal plates that are punched to create little spikes. Workers place them over the wood joint and push them into place with hydraulic presses.

9.1.6 Timber Rivets

Timber rivets have become popular since their introduction in the 1997 *NDS*[1]. Engineers developed timber rivet connections for glue-laminated timber, but they can be used in heavy timbers just as effectively. They are rectangular, hardened steel nails, typically 2½ in (65 mm) long. They install through ¼ in (6.5 mm) steel plates with round, drilled holes on a 1 in (25 mm) grid (Figure 9.17). The rivet heads bind in the steel plate hole, creating a tight, rigid connection. Timber rivets transfer load through bearing between the wood and rivet, then shear into the metal plate.

9.1.7 Split Ring and Shear Plate

We have utilized split ring and shear plate connectors since the 1940s, and they are still available today. They collect force from weaker wood over a

Figure 9.17 Timber rivets
Source: Photo courtesy of Gary C. Williams ©

large area, transfer it to a smaller steel area, and then move it back to wood over a large area. This has advantages over bolts alone. Split rings are rolled plate inserted into a groove cut in the connected members, with a bolt to hold the wood faces together, as seen in Figure 9.18. Force transfers from wood to steel ring to wood again. Shear plates are installed in a round, recessed portion milled into the connected members with a special dapping tool (Figure 9.19). Force is transferred from wood into the shear plate through the bolt, and into the other shear plate and member.

9.2 CAPACITY

Dowel-type connectors are loaded in two primary ways: lateral and withdrawal. Lateral loads place the connector in shear and the wood in bearing (see Figure 9.20a). The commonest are nails, bolts, and lag screws. Withdrawal places the connector in tension, essentially pulling it out of the wood (see Figure 9.20b). Nails and lag screws are often loaded in withdrawal.

Connector capacities are defined by the *NDS* in a similar fashion to wood materials. The primary difference is that the values are in units of force, not stress. Refer to Table 2.8 from Chapter 2 for the applicable

Figure 9.18 Split ring connector

Source: Photo courtesy of the Portland Bolt and Manufacturing Company ©

Paul W. McMullin

Figure 9.19 Shear plate connector

Source: Photo courtesy of the Portland Bolt and Manufacturing Company ©

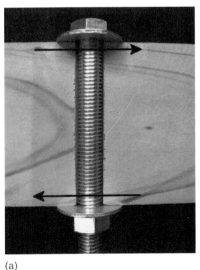
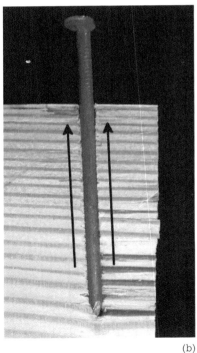

(a)

(b)

Figure 9.20 Connector load modes: (a) lateral (shear), and (b) withdrawal (tension)

Table 9.2 Fastener reference design strength summary of Appendix 3

Table	Description	Image
Table A3.2	Lag screw reference withdrawal values, W	
Table A3.1	Nail and spike reference withdrawal values, W	
Table A3.3	Dowel bearing strength, F_e	
Table A3.6	Bolt, single shear, all wood reference lateral design values, Z	
Table A3.7	Bolt, single shear, and steel plate side member reference lateral design values, Z	
Table A3.8	Bolt, double shear, all wood reference lateral design values, Z	

Paul W. McMullin

Table A3.9	Bolt, double shear, steel plate side member reference lateral design values, Z	
Table A3.10	Lag screw, single shear, all wood, reference lateral design values, Z	
Table A3.4	Common, box, sinker nail single shear, all wood, reference lateral design values, Z	
Table A3.5	Common, box, sinker nail single shear, steel side plate, reference lateral design values, Z	
Tables A3.11–A3.14	Selected Simpson™ connector reference strengths	

adjustment factors. Connector reference design strengths are a function of wood **specific gravity** (think density), rather than species. Appendix 3 contains connector reference design strengths; its contents are summarized in Table 9.2.

For bolted connections, we check local stresses in the wood to ensure strips of wood will not pull out, as illustrated in Figure 9.21. Find equations for this failure mode in Appendix E of the *NDS*[2].

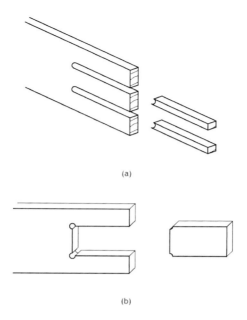

(a)

(b)

Figure 9.21 Local wood failure showing (a) multiple strip failure, and (b) group failure

9.2.1 Adjustment Factors

The adjustment factors for timber connectors are:

- load duration, C_D;
- wet service, C_M;
- temperature, C_t;
- group action, C_g;
- geometry, C_Δ;
- end grain, C_{eg};
- diaphragm, C_{di};
- toe-nail, C_{tn};
- LRFD conversion factors when appropriate, K_F, ϕ, λ.

Refer to the discussion of these factors in Section 2.4. See Table 2.8 in Chapter 2 to understand which adjustment factors apply to connector lateral and withdrawal action, and see Appendix 4 for their values. The geometry adjustment factor warrants additional discussion in this chapter.

Paul W. McMullin

9.2.1.a Geometry, C_Δ

The *NDS* provides minimum end and edge distance, and spacing requirements. This is to reduce the possibility of splitting and to engage the full fastener capacity. The requirements are provided in two groups: minimums for a reduced strength ($C_\Delta < 1.0$) and minimums for full strength ($C_\Delta = 1.0$). It is best to use at least the minimum required edge, **end distances**, and spacings in Section 9.4, so that the geometry factors are 1.0. When end distance and spacing between fasteners in a row are less than optimum, we calculate the geometry factor using the following equations:

$$C_\Delta = \frac{\text{actual end distance}}{\text{end distance for } C_\Delta = 1.0} \tag{9.1}$$

$$C_\Delta = \frac{\text{actual spacing in row}}{\text{spacing in row for } C_\Delta = 1.0} \tag{9.2}$$

Note that, for fastener diameters $D < \frac{1}{4}$ in (6.4 mm), the geometry factor $C_\Delta = 1.0$, which applies to most nail and spike diameters.

9.2.1.b Diaphragm, Cdi

When nails are used in diaphragms (floor or roof sheathing), the reference design strength, Z, can be multiplied by the diaphragm factor, C_{di}, of 1.1.

9.2.2 Engineered Metal Plate Connectors

Manufacturers publish allowable strengths for engineered metal plate connectors, or, better yet, they can be taken from an ICC ESR report. These reports are prepared by the International Code Council after it has approved the manufacturers' test data for the intended use. It sometimes reduces the strength values, or limits the connector's applicability for some specific building code requirements. Refer to Tables A3.11–A3.14 in Appendix 3 for allowable strengths of a few selected Simpson™ connectors. Consult Simpson's catalog for other connectors and the appropriate ICC ESR report for allowable strengths and limits on applicability.

The load duration, wet service, and temperature adjustment factors in Table 2.8 in Chapter 2 apply to engineered metal plate connectors. The factors related to geometry (C_g, C_Δ, C_{eg}, C_{di}, C_{tn}) are captured in the testing of a specific configuration. However, for proprietary nails or screws, we must consider all the adjustment factors. To convert engineered metal plate connector ASD values to LRFD, follow ASTM D5457[3].

9.3 DEMAND VS. CAPACITY

To determine if a connector is strong enough, we compare the required capacity with the adjusted design strength. We can check a connection on a total force basis or an individual fastener (i.e., a connector force) basis.

Often, it is easiest to divide the total force by the allowable force in a single fastener to determine the number, n, of required fasteners (n = force/strength). When the group adjustment factor is required, we can assume an initial value (say 0.75) to estimate the number, or just add a few more connectors. To close the loop, we calculate the group adjustment factor, apply this to the strength, and compare with load.

9.4 INSTALLATION

9.4.1 Spacing of Fasteners

Fastener spacing is of fundamental concern so that splitting of wood or failure of fasteners at lower loads than we anticipate can be avoided. We consider spacing between **rows of fasteners** (gage), spacing between fasteners in a row (pitch), end distance, and edge distance. These are defined in Figure 9.22. A row of fasteners is parallel to the direction of force. Dimensions are taken to the centerline of the fastener.

It is better to use more fasteners of a smaller diameter than a few large-diameter fasteners.

There are no code spacing or end and edge distance requirements for nails or spikes smaller than ¼ in (6.4 mm), which are intended to prevent wood splitting. However, as an aid, the values in Table 9.3 can be followed, as adopted from the *NDS Commentary*[4].

For dowel-type fasteners with diameters greater than ¼ in (6.4 mm), Tables 9.4–9.8 apply.

For design and detailing simplicity, Table 9.8 presents the distance corresponding to various dowel diameters.

Paul W. McMullin

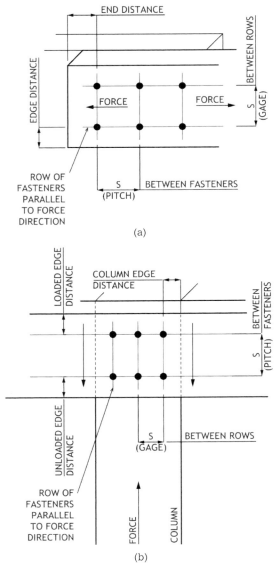

Figure 9.22 Fastener spacing, end, and edge distance requirements for load (a) parallel to grain, and (b) perpendicular to grain

Timber Connections

Table 9.3 Recommended nail spacing

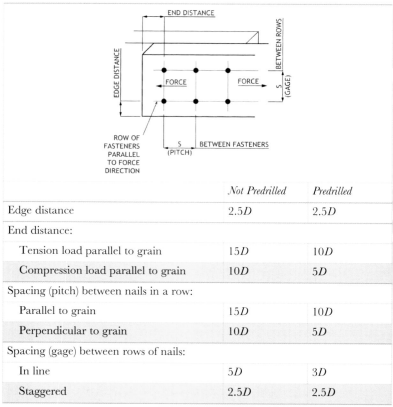

	Not Predrilled	*Predrilled*
Edge distance	2.5D	2.5D
End distance:		
Tension load parallel to grain	15D	10D
Compression load parallel to grain	10D	5D
Spacing (pitch) between nails in a row:		
Parallel to grain	15D	10D
Perpendicular to grain	10D	5D
Spacing (gage) between rows of nails:		
In line	5D	3D
Staggered	2.5D	2.5D

Notes: The code only requires that nails are spaced to not split the wood. These values are a guideline only. Species, moisture content, and grain orientation affect spacing

Source: NDS Commentary 2005

Paul W. McMullin

Table 9.4 Recommended nail spacing

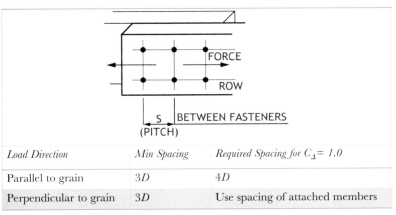

Load Direction	Min Spacing	Required Spacing for $C_\Delta = 1.0$
Parallel to grain	$3D$	$4D$
Perpendicular to grain	$3D$	Use spacing of attached members

Note: For $D <$ ¼ in (6.4 mm), $C_\Delta = 1.0$
Source: NDS 2015

Table 9.5 Spacing between rows of fasteners

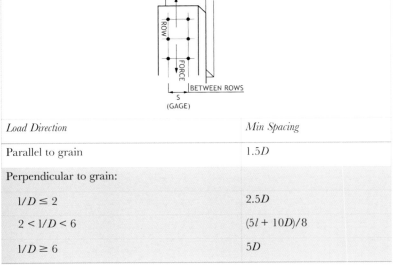

Load Direction	Min Spacing
Parallel to grain	$1.5D$
Perpendicular to grain:	
$l/D \leq 2$	$2.5D$
$2 < l/D < 6$	$(5l + 10D)/8$
$l/D \geq 6$	$5D$

Source: NDS 2015

Table 9.6 Fastener end distance requirements

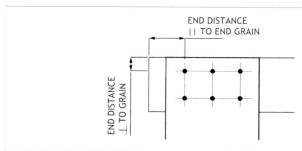

Load Direction	Min End Distance for $C_\Delta = 0.5$	Min End Distance for $C_\Delta = 1.0$
Perpendicular to grain	$2D$	$4D$
Parallel to grain—compression	$2D$	$4D$
Parallel to grain—tension:		
Softwoods	$3.5D$	$7D$
Hardwoods	$2.5D$	$5D$

Note: For $D < ¼$ in (6.4 mm), $C_\Delta = 1.0$
Source: NDS 2015

Table 9.7 Fastener edge distance requirements

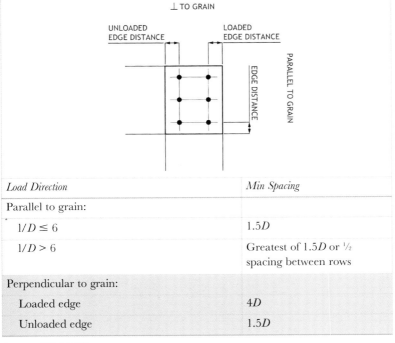

Load Direction	Min Spacing
Parallel to grain:	
$l/D \leq 6$	$1.5D$
$l/D > 6$	Greatest of $1.5D$ or $\frac{1}{2}$ spacing between rows
Perpendicular to grain:	
Loaded edge	$4D$
Unloaded edge	$1.5D$

Source: NDS 2015

Table 9.8 Spacing, end, or edge distances for given dowel diameter

Imperial Units

Fastener Diameter (in)	Spacing, End, or Edge Distances (in)									
	1.5D	2.0D	2.5D	3.0D	3.5D	4.0D	5.0D	7.0D	10.0D	15.0D
0.099	0.149	0.198	0.248	0.297	0.347	0.396	0.495	0.693	0.990	1.49
0.113	0.170	0.226	0.283	0.339	0.396	0.452	0.565	0.791	1.13	1.70
0.120	0.180	0.240	0.300	0.360	0.420	0.480	0.600	0.840	1.20	1.80
0.128	0.192	0.256	0.320	0.384	0.448	0.512	0.640	0.896	1.28	1.92
0.131	0.197	0.262	0.328	0.393	0.459	0.524	0.655	0.917	1.31	1.97
0.135	0.203	0.270	0.338	0.405	0.473	0.540	0.675	0.945	1.35	2.03
0.148	0.222	0.296	0.370	0.444	0.518	0.592	0.740	1.04	1.48	2.22
0.162	0.243	0.324	0.405	0.486	0.567	0.648	0.810	1.13	1.62	2.43
0.177	0.266	0.354	0.443	0.531	0.620	0.708	0.885	1.24	1.77	2.66
$\frac{1}{4}$	0.375	0.500	0.625	0.750	0.875	1.000	1.25	1.75	2.50	3.75
$\frac{5}{16}$	0.469	0.625	0.781	0.938	1.09	1.25	1.56	2.19	3.13	4.69
$\frac{3}{8}$	0.563	0.750	0.938	1.13	1.31	1.50	1.88	2.63	3.75	5.63
$\frac{7}{16}$	0.656	0.875	1.09	1.31	1.53	1.75	2.19	3.06	4.38	6.56
$\frac{1}{2}$	0.750	1.00	1.25	1.50	1.75	2.00	2.50	3.50	5.00	7.50
$\frac{9}{16}$	0.844	1.13	1.41	1.69	1.97	2.25	2.81	3.94	5.63	8.44
$\frac{5}{8}$	0.938	1.25	1.56	1.88	2.19	2.50	3.13	4.38	6.25	9.38
$\frac{3}{4}$	1.13	1.50	1.88	2.25	2.63	3.00	3.75	5.25	7.50	11.25
$\frac{7}{8}$	1.31	1.75	2.19	2.63	3.06	3.50	4.38	6.13	8.75	13.13
1	1.50	2.00	2.50	3.00	3.50	4.00	5.00	7.00	10.00	15.00

Paul W. McMullin

Metric Units

Fastener Diameter (mm)

Spacing, End, or Edge Distances (mm)

Fastener Diameter (mm)	1.5D	2.0D	2.5D	3.0D	3.5D	4.0D	5.0D	7.0D	10.0D	15.0D
2.51	3.77	5.03	6.29	7.544	8.801	10.1	12.6	17.6	25.1	37.7
2.87	4.31	5.74	7.18	8.611	10.0	11.5	14.4	20.1	28.7	43.1
3.05	4.57	6.10	7.62	9.144	10.7	12.2	15.2	21.3	30.5	45.7
3.25	4.88	6.50	8.13	9.754	11.4	13.0	16.3	22.8	32.5	48.8
3.33	4.99	6.65	8.32	9.982	11.6	13.3	16.6	23.3	33.3	49.9
3.43	5.14	6.86	8.57	10.3	12.0	13.7	17.1	24.0	34.3	51.4
3.76	5.64	7.52	9.40	11.3	13.2	15.0	18.8	26.3	37.6	56.4
4.11	6.17	8.23	10.3	12.3	14.4	16.5	20.6	28.8	41.1	61.7
4.50	6.74	8.99	11.2	13.5	15.7	18.0	22.5	31.5	45.0	67.4
6.35	9.53	12.7	15.9	19.1	22.2	25.4	31.8	44.5	63.5	95.3
7.94	11.9	15.9	19.8	23.8	27.8	31.8	39.7	55.6	79.4	119
9.53	14.3	19.1	23.8	28.58	33.34	38.1	47.6	66.7	95.3	143
11.1	16.7	22.2	27.8	33.3	38.9	44.5	55.6	77.8	111	167
12.7	19.1	25.4	31.8	38.1	44.5	50.8	63.5	88.9	127	191
14.3	21.4	28.6	35.7	42.9	50.0	57.2	71.4	100	143	214
15.9	23.8	31.8	39.7	47.6	55.6	63.5	79.4	111	159	238
19.1	28.6	38.1	47.6	57.2	66.7	76.2	95.3	133	191	286
22.2	33.3	44.5	55.6	66.7	77.8	88.9	111	156	222	333
25.4	38.1	50.8	63.5	76.2	88.9	102	127	178	254	381

Timber Connections

9.4.2 Penetration

Penetration of nails and lag screws into the main member is fundamental to their ability to carry loads. The *NDS* requires the following minimum penetration, P_{min}, as illustrated in Figure 9.23:

- nails and spikes—six times the diameter ($6D$), including the tip;
- lag screws—four times the diameter ($4D$), not including the tip.

9.4.3 Minimum Nailing

To ensure adequate connections, the International Building Code[5] requires minimum nailing between members, given in Table 9.9.

9.4.4 Preboring

Preboring, or predrilling, provides space for dowel-type fasteners in the wood, thereby reducing splitting. For bolts, the hole is slightly larger than

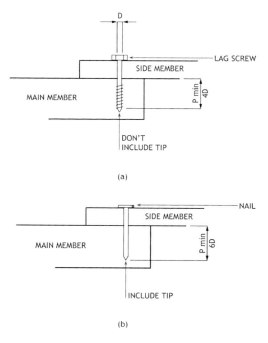

(a)

(b)

Figure 9.23 (a) Nail and (b) lag screw minimum penetration

Paul W. McMullin

Table 9.9 Selected minumum nailing requirements

Connection	Nailing
Joist to sill or girder	(3) 8d toe-nail
Bridging to joist	(2) 8d toe-nail
Top plate to stud	(2) 16d endnail
Stud to sole plate	(2) 16d endnail
Blocking to top plate	(3) 8d toe-nail
Rim joist to top plate	8d @ 16 in (150 mm) toe-nail
Top plate laps	(2) 16d face-nail
Multilayer header	16d @ 16 in (150 mm) face-nail
Rafter to top plate	(3) 8d toe-nail
Built-up corner studs	16d @ 24 in (600 mm) face-nail
Collar tie to rafter	(3) 10d face-nail
Rafter to 2× ridge beam	(2) 10d face-nail

Source: IBC 2012

the shaft diameter. For lag screws and nails, the holes are smaller by a percentage of the fastener diameter.

For nails, preboring recommendations are no longer in the code or commentary. Utilizing past requirements[6], we can use the following percentages of the nail diameter:

- 75 percent for $G \leq 0.59$;
- 90 percent for $G > 0.59$.

Holes for bolts should be a minimum of $\frac{1}{32}$ in (0.8 mm) and a maximum of $\frac{1}{16}$ in (1.8 mm) greater than the bolt diameter.

For lag screws, we choose a preboring size to prevent splitting. Use the bolt size requirements for the side member hole. For the lead hole (where threads engage), follow these recommendations:

- 40–70 percent of nominal screw size for $G \leq 0.5$;
- 60–75 percent of nominal screw size for $0.5 < G \leq 0.6$;
- 65–85 percent of nominal screw size for $G > 0.6$.

9.5 DETAILING

This section provides detail examples of the following timber connections:

- joist to beam—Figure 9.24;
- beam to column—Figure 9.25;
- light truss to bearing wall—Figure 9.26;
- floor to wall—Figure 9.27;
- header on **jamb studs**—Figure 9.28;
- column base—Figure 9.29;
- shearwall hold down—Figure 9.30;
- timber truss joint—Figure 9.31.

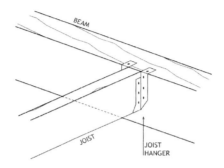

Figure 9.24 Joist to beam connection

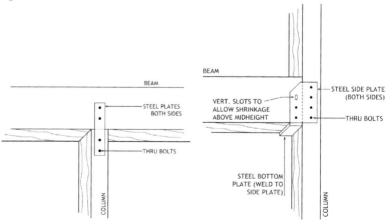

Figure 9.25 Beam–column connection

Paul W. McMullin

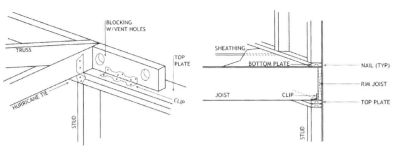

Figure 9.26 Truss to stud wall connection

Figure 9.27 Floor joist to stud wall connection

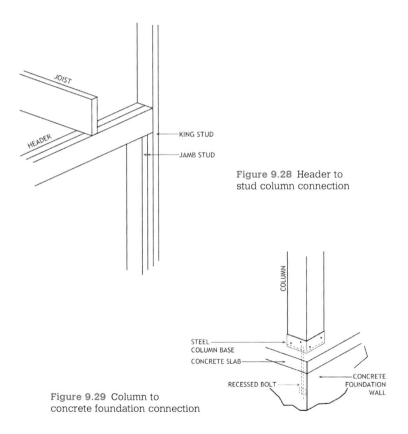

Figure 9.28 Header to stud column connection

Figure 9.29 Column to concrete foundation connection

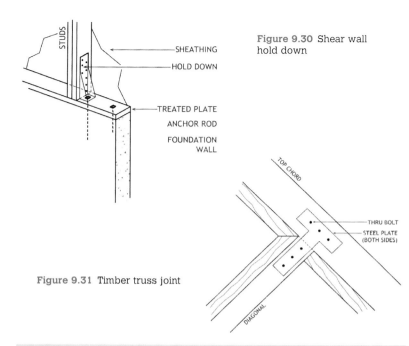

Figure 9.30 Shear wall hold down

STUDS
SHEATHING
HOLD DOWN
TREATED PLATE
ANCHOR ROD
FOUNDATION WALL

TOP CHORD
THRU BOLT
STEEL PLATE (BOTH SIDES)
DIAGONAL

Figure 9.31 Timber truss joint

9.6 DESIGN STEPS

Follow these steps when designing a connection:

1. Determine the force in the connector group and/or individual connector.
2. Choose the fastener type. This decision is a function of load magnitude, constructability, contractor preference, and aesthetics.
3. Find the connector reference design strength from the applicable table in Appendix 3.
4. Determine which adjustment factors apply (use Table 2.8) and find them, utilizing Section 2.4, Table 2.3, and Appendix 4.
5. Multiply the reference design strength by the adjustment factors to find the adjusted design strength.
6. Compare the strength with the connection load.
7. Sketch the final connection geometry.

Paul W. McMullin

9.7 DESIGN EXAMPLES

We explore three types of connection in this section: truss bottom chord in tension, a beam to column connection, and a joist hanger, illustrated in Figure 9.32.

9.7.1 Truss Chord Repair Example

This example is a bottom truss chord in tension. It needs a field repair after someone cut through it while installing a dryer vent.

Step 1: Calculate Force in Connector Group

We have calculated the force to be:

$T = 1,500$ lb $\qquad\qquad T = 6,672$ N

(See Chapter 7 for additional guidance on truss analysis.)

Step 2: Choose Fastener Type

For ease of installation and to limit splitting, we will use 8d **box nails**. We will use ¾ in plywood on each side of the cut, shown in Figure 9.33.

Step 3: Find Fastener Reference Design Strength

Go to Appendix 3, Table A3.4.

We find the reference design strength by entering the table and finding the side member thickness of ¾ in (19 mm). We then go over to

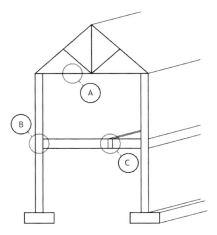

Figure 9.32 Example structure view showing connection locations

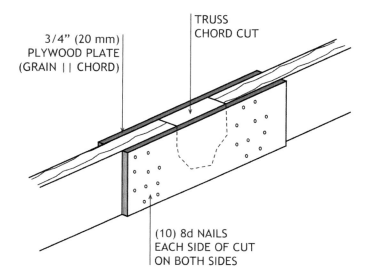

TRUSS
CHORD CUT

3/4" (20 mm)
PLYWOOD PLATE
(GRAIN || CHORD)

(10) 8d NAILS
EACH SIDE OF CUT
ON BOTH SIDES

Figure 9.33 Example truss bottom chord splice

the column that says 'Box' and go down to the row labeled 8d.
Going to the right, we find the column under Douglas Fir–Larch (N) and
read:

Z = 71 lb Z = 316 N

A straight edge helps keep the rows and columns straight.

Step 4: Determine Adjustment Factors

From Table 2.8, in Chapter 2, we see which adjustment factors apply to
nails (dowel-type fasteners) under lateral loads. Following Table 2.3 and
Appendix 4, we will find each factor in the table.

As complicated as it seems at first, many of the adjustment factors
frequently don't apply, and are therefore 1.0. Take your time and it will
become clear.

Factor	Description	Source
$C_D = 1.15$	Load duration—snow loads govern, so we get a slight advantage	Table A4.1
$C_M = 1.0$	Wet service—dry, interior condition	Table A4.2
$C_t = 1.0$	Temperature—sustained temperatures don't exceed 100°F (37.8°C)	Table A4.4
$C_G = 1.0$	Group action—does not apply to nails	Table A4.12
$C_\Delta = 1.0$	Geometry—does not apply to diameters under $\frac{1}{4}$ in (6 mm)	Tables A4.13 & A4.14
$C_{eg} = 1.0$	End grain—nails are installed on side face	Table A4.15
$C_{di} = 1.0$	Diaphragm—this is not a diaphragm	
$C_{tn} = 1.0$	Toe-nail—nails are not installed in toe-nail fashion	Table A4.16

Step 5: Multiply Reference Design Strength by Adjustment Factors

To find the adjusted design strength, we multiply the reference design strength by the adjustment factors to get:

$$Z' = Z C_D C_M C_t C_g C_\Delta C_{eg} C_{di} C_{tn}$$

$Z' = 71$ lb $(1.15)(1.0)$ $\qquad$ $Z' = 316\ N\,(1.15)(1.0)$

$Z' = 81.7$ lb $\qquad\qquad\qquad$ $Z' = 363.4\ N$

Step 6: Compare Strength with Load

We have calculated the strength of one nail above. We find the number of required nails by dividing the demand by capacity, as follows:

$$n = \frac{T}{Z'}$$

$$\frac{1{,}500\ \text{lb}}{81.7\ \text{lb/nail}} = 18.4 \text{ nails} \qquad \frac{6{,}672\ \text{N}}{363.4\ \text{N/nail}} = 18.4 \text{ nails}$$

We will round up to 20 nails. This is the total number of fasteners on each side of the cut. Because we are using two side plates, we will place ten nails on each side of the cut, into each side member, for a total of 40 nails.

Step 7: Sketch the Connection

Figure 9.33 shows the final connection.

9.7.2 Beam–Column Connection Example

We now look at a beam–column connection, where the column continues past the beam. The column and beam are different species, as we are using recycled wood.

Step 1: Calculate Force in Connector Group

From the beam analysis, we know the force at the end is:

R = 19,500 lb $\qquad\qquad$ R = 86.74 kN

Step 2: Choose Fastener Type

We will use ¾ in (19 mm) diameter bolts for their strength compared with nails. We will use ¼ in (6.4 mm) steel side plates, shown in Figure 9.34. The column is a 6 in (150 mm) square.

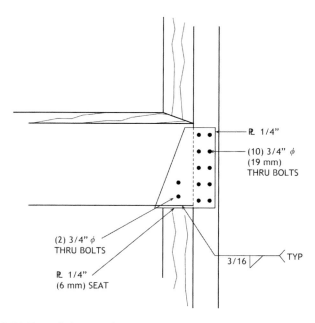

Figure 9.34 Example beam–column connection

d = 5.5 in d = 150 mm

b = 5.5 in b = 150 mm

$A_m = bd$

= 5.5 in (5.5 in) = 30.25 in²   = 150 mm (150 mm) = 22,500 mm²

Step 3: Find Fastener Reference Design Strength

Going to Appendix 3, Table A3.9, we find the reference design strength by entering the table and finding the main member thickness of 5¼ in (133 mm). Finding the ¼ in (6.4 mm) row, we go right to the Douglas Fir–Larch (N) column. We read off both the $Z_\parallel$ and $Z_\perp$ values:

$Z_\parallel$ 3,320 lb $Z_\parallel$ 14.77 kN

$Z_\perp$ 1,850 lb $Z_\perp$ 8.23 kN

Step 4: Determine Adjustment Factors

Using Table 2.8, in Chapter 2, we see which adjustment factors apply to bolts under lateral loads. Following Table 2.3 and Appendix 4, we will find each factor shown in the table.

Factor	Description	Source
C_D = 1.0	Load duration—no increase for floor load	Table A4.1
C_M = 1.0	Wet service—dry, interior condition	Table A4.2
C_t = 1.0	Temperature—regular temperature	Table A4.4
C_Δ= 1.0	Geometry—use spacing, end, and edge distances so we don't have a reduction	Tables A4.13 & A4.14
C_{eg} = 1.0	End grain—bolts are installed on side face	Table A4.15
C_{di} = 1.0	Diaphragm—this is not a diaphragm	
C_{tn} = 1.0	Toe-nail—bolts are not installed in toe-nail fashion	Table A4.16

To find the group adjustment factor, C_g, we will make a quick estimate of the number of required bolts by dividing the force by reference design strength.

$$n = \frac{R}{Z_{ll}}$$

$$= \frac{19{,}500 \text{ lb}}{3{,}320 \text{ lb/bolt}} = 5.9 \qquad = \frac{86.74 \text{ kN}}{14.77 \text{ kN/bolt}} = 5.9$$

We will use $n = 8$ to account for the group factor and give some extra capacity. With two rows of bolts, we have four bolts in a row. From Table A4.12, we find $C_g = 0.96$.

Step 5: Multiply the Reference Design Strength by Adjustment Factors

$Z_\parallel = Z_\parallel C_D C_M C_t C_g C_\Delta C_{eg} C_{di} C_{tn}$

$= 3{,}320 \text{ lb } (1.0) \ 0.96 \ (1.0) \qquad = 14.77 \text{ kN } (1.0) \ 0.96 \ (1.0)$

$= 3{,}187 \text{ lb} \qquad\qquad\qquad\qquad = 14.18 \text{ kN}$

Step 6: Compare Strength with Load

We will look at strength on a connection basis by multiplying a single bolt strength by the number of bolts:

$Z'_{\text{conn}} = nZ'$

$= 8 \ (3{,}187 \text{ lb}) = 25{,}498 \text{ lb} \qquad = 8 \ (14.18 \text{ kN}) = 113.4 \text{ kN}$

Because this is greater than the reaction, we are OK.

For the bolts between the beam and steel plate, we would need roughly twice the number of bolts, because the force is perpendicular to the grain. However, this creates cross-grain tension when the bolts are above the mid-height of the beam. To avoid this, we will use a bearing plate between the two side plates to support the beam. The bolts in the beam will just be to keep it from sliding off its bearing seat.

To size the bearing seat, we need the bearing strength perpendicular to the grain. From Table A2.4 in Appendix 2, and selecting Southern Pine No.1, we have:

$F_{c\perp} = 565 \text{ lb/in}^2 \qquad F_{c\perp} = 3{,}896 \text{ kN/m}^2$

We will use the same C_M and C_t factors from above. The remaining two factors follow the table.

The adjusted design strength is

$F'_{c\perp} = F'_{c\perp} C_M C_t C_i C_b$

$= 565 \text{ lb/in}^2 \ (1.0) = 565 \text{ lb/in}^2 \qquad = 3{,}896 \text{ kN/m}^2 \ (1.0) = 3{,}896 \text{ kN/m}^2$

Factor	Description	Source
$C_i = 1.0$	Incising—the timber is not pressure treated	Table A4.10
$C_b = 1.0$	Bearing—load is applied at the end	Table A4.11

To find the required bearing area, we divide the reaction by bearing strength:

$$A_{brg} = \frac{R}{F'_{c\perp}}$$

$$= \frac{19,500\ \text{lb}}{565\ \text{lb/in}^2}$$

$$= 34.5\ \text{in}^2$$

$$= \frac{86.74\ \text{kN}}{3,896\ \text{kN/m}^2}\ \frac{1000^2\ \text{mm}^2}{1\ \text{m}^2}$$

$$= 22,264\ \text{mm}^2$$

Dividing beam area by width, b_{bm}, we get the necessary bearing length:

$$b_{bm} = 5.5\ \text{in} \qquad\qquad b_{bm} = 140\ \text{mm}$$

$$l_{brg} = \frac{A_{brg}}{b_{bm}}$$

$$= \frac{34.5\ \text{in}^2}{5.5\ \text{in}}$$

$$= 6.3\ \text{in}$$

$$= \frac{22,264\ \text{mm}^2}{140\ \text{mm}}$$

$$= 159\ \text{mm}$$

Step 7: Sketch the Connection

Figure 9.34 shows the final connection. We add an extra set of bolts to account for the **eccentricity** of the bearing seat.

9.7.3 Joist Hanger Connection Example

We now briefly look at a joist–beam connection.

Step 1: Calculate Force in Connector

We calculate the end reaction of the joist, assuming a uniform, distributed load and the following:

$$l = 20\ \text{ft} \qquad\qquad l = 6.10\ \text{m}$$

$$s = 16\ \text{in} \qquad\qquad s = 400\ \text{mm}$$

D = 20 lb/ft² D = 960 N/m²

L = 50 lb/ft² L = 2,390 N/m²

$$R = (D+L)\frac{l}{2}s$$

$$= \left(20\frac{\text{lb}}{\text{ft}^2} + 50\frac{\text{lb}}{\text{ft}^2}\right)\frac{20 \text{ ft}}{2} \, 16 \text{ in} \frac{1 \text{ ft}}{12 \text{ in}} \qquad = \left(960\frac{\text{N}}{\text{m}^2} + 2,390\frac{\text{N}}{\text{m}^2}\right)\frac{6.10 \text{ m}}{2} \, 0.4 \text{ m}$$

$$= 933 \text{ lb} \qquad\qquad\qquad\qquad\qquad = 4,087 \text{ N}$$

Step 2: Choose Fastener Type

The joist is a 11⅞ in (302 mm) I-joist with a 1¾ in (45 mm) wide flange, shown in Figure 9.35. Let's try an ITS2.06/11.88 Simpson hanger.

Step 3: Find Fastener Reference Design Strength

Going to Appendix 3, Table A3.12, we see:

Z_S = 1,150 lb Z_S = 5,115 N

Step 4: Determine Adjustment Factors

Using Table 2.8, in Chapter 2, we see which adjustment factors apply. Following Table 2.3 and Appendix 3, we will find each factor in the table.

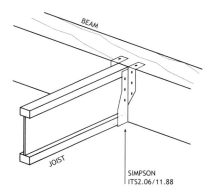

Figure 9.35 Example joist hanger to beam connection

Paul W. McMullin

Factor	Description	Source
$C_D = 1.0$	Load duration—no increase for floor load	Table A4.1
$C_M = 1.0$	Wet service—dry, interior condition	Table A4.2
$C_t = 1.0$	Temperature—regular temperature	Table A4.4

Step 5: Multiply Reference Design Strength by Adjustment Factors

To find the adjusted design strength, we multiply the reference design strength by the adjustment factors to get:

$Z'_s = Z_s C_D C_M C_t$

$= 1{,}150\ \text{lb}\ (1.0) = 1{,}150\ \text{lb}$ $= 5{,}115\ \text{N}\ (1.0) = 5{,}115\ \text{N}$

Step 6: Compare Strength with Load

Because the connector strength is greater than the joist reaction, we are OK.

Step 7: Sketch the Connection

Figure 9.35 shows the final connection.

9.8 RULES OF THUMB

A simple rule of thumb is to take a lower-bound strength for each connector type. We can find the number of required fasteners in a joint by dividing the required force by this value. Table 9.10 presents lower-bound strength values for nails, bolts, and lag screws. They are a good starting point to determine the number of required fasteners in a group. For example, say we have a nailed connection with a 2,000 lb (8,900 N) lateral load through it. We choose a 10d nail and divide its lower-bound strength into the load:

$n = 2{,}000\ \text{lb}/110\ \text{lb/nail}$ $n = 8{,}900\ \text{N}/490\ \text{N/nail}$

$= 18\ \text{nails}$ $= 18\ \text{nails}$

Table 9.10 Lower-bound rule of thumb dowel fastener strength

Fastener Type	Size	Strength[1,2,3]					
		Lateral				Withdrawal[5]	
		Perpendicular to Grain		Parallel to Grain			
		Imperial (lb)	Metric (N)	Imperial (lb)	Metric (N)	Imperial (lb)	Metric (N)
Nails[4]	8d	90	(400)	90	(400)	20	(90)
	10d	110	(490)	110	(490)	30	(130)
	16d	140	(620)	140	(620)	35	(150)
	40d	200	(890)	200	(890)	70	(310)
Bolts	$\frac{1}{2}$	610	(2,800)	500	(2,200)	not applicable	
	$\frac{5}{8}$	870	(3,900)	600	(2,700)		
	$\frac{3}{4}$	1,190	(5,300)	700	(3,100)		
	$\frac{7}{8}$	1,570	(7,000)	800	(3,600)		
	1	1,790	(8,000)	950	(4,200)		
Lag screws	$\frac{1}{4}$	110	(490)	150	(670)	200	(890)
	$\frac{5}{16}$	130	(580)	200	(890)	300	(1,300)
	$\frac{3}{8}$	140	(620)	200	(890)	400	(1,800)
	$\frac{1}{2}$	220	(980)	390	(1,700)	700	(3,100)
	$\frac{5}{8}$	310	(1,400)	550	(2,400)	1,000	(4,400)
	$\frac{3}{4}$	430	(1,900)	760	(3,300)	1,400	(6,200)

Notes: (1.) These values are lower bound to give an idea of the number of required fasteners for a given load. Use actual values from Appendix 3 for design. (2.) Values based on $G = 0.49$; multiply by ratio of $Gactual/G$ for other specific gravities. (3.) Values based on 1.5 in (38 mm) wood side member and 3½ in (90 mm) main members. (4.) Values based on common nails. (5.) Values based on $p_{min} = 6D$ for nails, and $p_{min} = 4D$ for lag screws

Paul W. McMullin

Where conditions exist (load duration, moisture content, temperature) that will reduce the strengths in Table 9.10, apply these factors to the lower-bound values.

9.9 WHERE WE GO FROM HERE

The number of available timber connectors is extensive. We have covered dowel-type fasteners and introduced a variety of other connectors. From here, we consider load, connection geometry, and connector strength to develop additional timber connections.

For connectors that are not covered in the *NDS*, it is important that they carry an ICC report. This ensures that an independent, code-body-affiliated group reviews the test data and limits their use when appropriate. It is wise to use the test data published in these reports, rather than the manufacturer's literature, as they may differ.

NOTES

1. AF&PA. *National Design Specification (NDS) for Wood Construction* (Washington, DC: American Forest & Paper Association, 1997).
2. ANSI/AWC. *National Design Specification (NDS) for Wood Construction Commentary* (Washington, DC: American Forest & Paper Association, 2015), p. 162.
3. ASTM. *Standard Specification for Computing Reference Resistance of Wood-Based Materials and Structural Connections for Load and Resistance Design* (West Conshohocken, PA; ASTM International, 2015).
4. AF&PA. *National Design Specification (NDS) for Wood Construction Commentary* (Washington DC: American Forest & Paper Association, 2005), p. 228.
5. IBC. *International Building Code* (Washington, DC: International Code Council, 2012).
6. AF&PA. *Commentary on the National Design Specification (NDS) for Wood Construction* (Washington, DC: American Forest & Paper Association, 1997), p. 140.

Section Properties

Appendix 1

Table A1.1 Sawn lumber section properties

Imperial Units

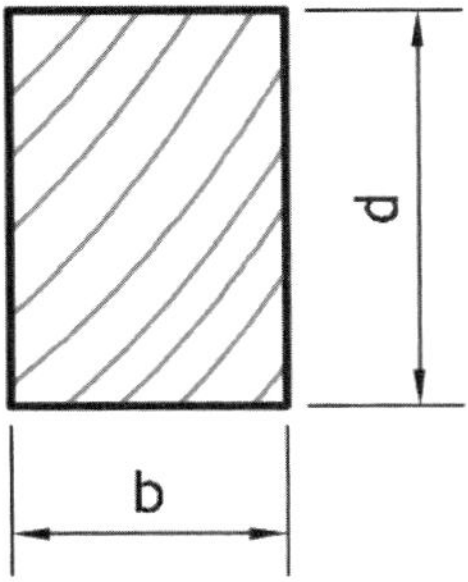

Nominal Size		Dressed Size			X–X Axis			Y–Y Axis		
b (in)	d (in)	b (in)	d (in)	A (in²)	S_{xx} (in³)	I_{xx} (in⁴)	r_{xx} (in)	S_{yy} (in³)	I_{yy} (in⁴)	r_{yy} (in)
Dimension Lumber										
2	4	1½	3½	5.250	3.060	5.359	1.010	1.313	0.984	0.433
	6		5½	8.250	7.560	20.80	1.588	2.063	1.547	0.433
	8		7¼	10.88	13.14	47.63	2.092	2.719	2.039	0.433
	10		9¼	13.88	21.39	98.93	2.670	3.469	2.602	0.433
	12		11¼	16.88	31.64	178.0	3.247	4.219	3.164	0.433
	14		13¼	19.88	43.89	290.8	3.825	4.969	3.727	0.433
4	4	3½	3½	12.25	7.150	12.51	1.011	7.146	12.51	1.011
	6		5½	19.25	17.65	48.53	1.588	11.23	19.65	1.010
	8		7¼	25.38	30.66	111.1	2.092	14.80	25.90	1.010
	10		9¼	32.38	49.91	230.8	2.670	18.89	33.05	1.010
	12		11¼	39.38	73.83	415.3	3.247	22.97	40.20	1.010
	14		13¼	46.38	102.4	678.5	3.825	27.05	47.34	1.010
	16		15¼	53.38	135.7	1,034	4.401	31.14	54.49	1.010

Nominal Size		Dressed Size		X–X Axis				Y–Y Axis		
b (in)	d (in)	b (in)	d (in)	A (in^2)	S_{xx} (in^3)	I_{xx} (in^4)	r_{xx} (in)	S_{yy} (in^3)	I_{yy} (in^4)	r_{yy} (in)
Columns										
5	5	$4^1/_2$	$4^1/_2$	20.25	15.19	34.17	1.299	15.19	34.17	1.299
6	6	$5^1/_2$	$5^1/_2$	30.25	27.73	76.26	1.588	27.73	76.26	1.588
	8		$7^1/_2$	41.25	51.56	193.4	2.165	37.81	104.0	1.588
8	8	$7^1/_2$	$7^1/_2$	56.25	70.31	263.7	2.165	70.31	263.7	2.165
	10		$9^1/_2$	71.25	112.8	535.9	2.742	89.06	334.0	2.165
10	10	$9^1/_2$	$9^1/_2$	90.25	142.9	678.8	2.742	142.9	678.8	2.742
	12		$11^1/_2$	109.3	209.4	1,204	3.320	173.0	821.7	2.742
12	12	$11^1/_2$	$11^1/_2$	132.3	253.5	1,458	3.320	253.5	1,458	3.320
	14		$13^1/_2$	155.3	349.3	2,358	3.897	297.6	1,711	3.320
14	14	$13^1/_2$	$13^1/_2$	182.3	410.1	2,768	3.897	410.1	2,768	3.897
	16		$15^1/_2$	209.3	540.6	4,189	4.474	470.8	3,178	3.897
16	16	$15^1/_2$	$15^1/_2$	240.3	620.6	4,810	4.474	620.6	4,810	4.474
	18		$17^1/_2$	271.3	791.1	6,923	5.052	700.7	5,431	4.474
18	18	$17^1/_2$	$17^1/_2$	306.3	893.2	7,816	5.052	893.2	7,816	5.052
	20		$19^1/_2$	341.3	1,109	10,813	5.629	995.3	8,709	5.052
20	20	$19^1/_2$	$19^1/_2$	380.3	1,236	12,049	5.629	1,236	12,049	5.629
	22		$21^1/_2$	419.3	1,502	16,150	6.207	1,363	13,285	5.629
22	22	$21^1/_2$	$21^1/_2$	462.3	1,656	17,806	6.207	1,656	17,806	6.207
	24		$23^1/_2$	505.3	1,979	23,252	6.784	1,810	19,463	6.207
24	24	$23^1/_2$	$23^1/_2$	552.3	2,163	25,415	6.784	2,163	25,415	6.784

Table A1.1 *continued*

Nominal Size		Dressed Size			X–X Axis			Y–Y Axis		
b (in)	d (in)	b (in)	d (in)	A (in²)	S_{xx} (in³)	I_{xx} (in⁴)	r_{xx} (in)	S_{yy} (in³)	I_{yy} (in⁴)	r_{yy} (in)
Beams										
6	10	5½	9½	52.25	82.73	393.0	2.742	47.90	131.7	1.588
	12	5½	11½	63.25	121.2	697.1	3.320	57.98	159.4	1.588
	14	5½	13½	74.25	167.1	1,128	3.897	68.06	187.2	1.588
	16	5½	15½	85.25	220.2	1,707	4.474	78.15	214.9	1.588
	18	5½	17½	96.25	280.7	2,456	5.052	88.23	242.6	1.588
	20	5½	19½	107.3	348.6	3,398	5.629	98.31	270.4	1.588
	22	5½	21½	118.3	423.7	4,555	6.207	108.4	298.1	1.588
	24	5½	23½	129.3	506.2	5,948	6.784	118.5	325.8	1.588
8	12	7½	11½	86.25	165.3	950.5	3.320	107.8	404.3	2.165
	14	7½	13½	101.3	227.8	1,538	3.897	126.6	474.6	2.165
	16	7½	15½	116.3	300.3	2,327	4.474	145.3	544.9	2.165
	18	7½	17½	131.3	382.8	3,350	5.052	164.1	615.2	2.165
	20	7½	19½	146.3	475.3	4,634	5.629	182.8	685.5	2.165
	22	7½	21½	161.3	577.8	6,211	6.207	201.6	755.9	2.165
	24	7½	23½	176.3	690.3	8,111	6.784	220.3	826.2	2.165
10	14	9½	13½	128.3	288.6	1,948	3.897	203.1	964.5	2.742
	16	9½	15½	147.3	380.4	2,948	4.474	233.1	1,107	2.742
	18	9½	17½	166.3	484.9	4,243	5.052	263.2	1,250	2.742
	20	9½	19½	185.3	602.1	5,870	5.629	293.3	1,393	2.742
	22	9½	21½	204.3	731.9	7,868	6.207	323.4	1,536	2.742
	24	9½	23½	223.3	874.4	10,274	6.784	353.5	1,679	2.742
12	16	11½	15½	178.3	460.5	3,569	4.474	341.6	1,964	3.320
	18	11½	17½	201.3	587.0	5,136	5.052	385.7	2,218	3.320
	20	11½	19½	224.3	728.8	7,106	5.629	429.8	2,471	3.320
	22	11½	21½	247.3	886.0	9,524	6.207	473.9	2,725	3.320
	24	11½	23½	270.3	1,058	12,437	6.784	518.0	2,978	3.320
14	18	13½	17½	236.3	689.1	6,029	5.052	531.6	3,588	3.897
	20	13½	19½	263.3	855.6	8,342	5.629	592.3	3,998	3.897
	22	13½	21½	290.3	1,040	11,181	6.207	653.1	4,408	3.897
	24	13½	23½	317.3	1,243	14,600	6.784	713.8	4,818	3.897

Table A1.1 *continued*

Nominal Size		Dressed Size			X–X Axis				Y–Y Axis		
b (in)	d (in)	b (in)	d (in)	A (in^2)	S_{xx} (in^3)	I_{xx} (in^4)	r_{xx} (in)	S_{yy} (in^3)	I_{yy} (in^4)	r_{yy} (in)
Beams											
16	20	15½	19½	302.3	982.3	9,578	5.629	780.8	6,051	4.474
	22	15½	21½	333.3	1,194	12,837	6.207	860.9	6,672	4.474
	24	15½	23½	364.3	1,427	16,763	6.784	941.0	7,293	4.474
18	22	17½	21½	376.3	1,348	14,493	6.207	1,097	9,602	5.052
	24	17½	23½	411.3	1,611	18,926	6.784	1,199	10,495	5.052
20	24	19½	23½	458.3	1,795	21,089	6.784	1,489	14,521	5.629

Appendix 1

Table A1.1 Sawn lumber section properties

Metric Units

Nominal Size		Dressed Size			X–X Axis			Y–Y Axis		
b	d	b	d	A	S_{xx} $\times 10^6$	I_{xx} $\times 10^6$	r_{xx}	S_{yy} $\times 10^6$	I_{yy} $\times 10^6$	r_{yy}
(mm)	(mm)	(mm)	(mm)	(mm²)	(mm³)	(mm⁴)	(mm)	(mm³)	(mm⁴)	(mm)
Dimension Lumber										
50.8	101.6	38	89	3,382	0.050	2.232	0.026	0.021	0.407	0.011
	152.4		140	5,320	0.124	8.689	0.040	0.034	0.640	0.011
	203.2		184	6,992	0.214	19.73	0.053	0.044	0.841	0.011
	254		235	8,930	0.350	41.10	0.068	0.057	1.075	0.011
	304.8		286	10,868	0.518	74.08	0.083	0.069	1.308	0.011
	355.6		337	12,806	0.719	121.2	0.097	0.081	1.541	0.011
101.6	101.6	89	89	7,921	0.117	5.229	0.026	0.117	5.229	0.026
	152.4		140	12,460	0.291	20.35	0.040	0.185	8.225	0.026
	203.2		184	16,376	0.502	46.20	0.053	0.243	10.81	0.026
	254		235	20,915	0.819	96.25	0.068	0.310	13.81	0.026
	304.8		286	25,454	1.213	173.5	0.083	0.378	16.80	0.026
	355.6		337	29,993	1.685	283.9	0.097	0.445	19.80	0.026
	406.4		387	34,443	2.222	429.9	0.112	0.511	22.74	0.026

Nominal Size		Dressed Size			X–X Axis			Y–Y Axis		
b	*d*	*b*	*d*	*A*	S_{xx} $\times 10^6$	I_{xx} $\times 10^6$	r_{xx}	S_{yy} $\times 10^6$	I_{yy} $\times 10^6$	r_{yy}
(mm)	(mm)	(mm)	(mm)	(mm²)	(mm³)	(mm⁴)	(mm)	(mm³)	(mm⁴)	(mm)
Columns										
127	127	114	114	12,996	0.247	14.07	0.03	0.247	14.07	0.033
152.4	152.4	139	140	19,460	0.454	31.78	0.04	0.451	31.33	0.040
	203.2		191	26,549	0.845	80.71	0.06	0.615	42.75	0.040
203.2	203.2	191	191	36,481	1.161	110.9	0.06	1.161	110.9	0.055
	254		241	46,031	1.849	222.8	0.07	1.465	139.9	0.055
254	254	241	241	58,081	2.333	281.1	0.07	2.333	281.1	0.070
	304.8		292	70,372	3.425	500.0	0.08	2.827	340.6	0.070
304.8	304.8	292	292	85,264	4.150	605.8	0.08	4.150	605.8	0.084
	355.6		343	100,156	5.726	981.9	0.10	4.874	711.6	0.084
355.6	355.6	343	343	117,649	6.726	1,153	0.10	6.726	1,153	0.099
	406.4		394	135,142	8.874	1,748	0.114	7.726	1,325	0.099
406.4	406.4	394	394	155,236	10.19	2,008	0.114	10.19	2,008	0.114
	457.2		445	175,330	13.00	2,893	0.128	11.51	2,268	0.114
457.2	457.2	445	445	198,025	14.69	3,268	0.128	14.69	3,268	0.128
	508		495	220,275	18.17	4,498	0.143	16.34	3,635	0.128
508	508	495	495	245,025	20.21	5,003	0.143	20.21	5,003	0.143
	558.8		546	270,270	24.59	6,714	0.158	22.30	5,519	0.143
558.8	558.8	546	546	298,116	27.13	7,406	0.158	27.13	7,406	0.158
	609.6		597	325,962	32.43	9,681	0.172	29.66	8,098	0.158
609.6	609.6	597	597	356,409	35.46	10,586	0.172	35.46	10,586	0.172

Table A1.1 *continued*

Nominal Size		*Dressed Size*			*X–X Axis*			*Y–Y Axis*		
b	*d*	*b*	*d*	*A*	S_{xx} $\times 10^6$	I_{xx} $\times 10^6$	r_{xx}	S_{yy} $\times 10^6$	I_{yy} $\times 10^6$	r_{yy}
(mm)	*(mm)*	*(mm)*	*(mm)*	*(mm²)*	*(mm³)*	*(mm⁴)*	*(mm)*	*(mm³)*	*(mm⁴)*	*(mm)*
Beams										
152.4	254	140	241	33,740	1.355	163.3	0.070	0.787	55.11	0.040
	304.8		292	40,880	1.989	290.5	0.084	0.954	66.77	0.040
	355.6		343	48,020	2.745	470.8	0.099	1.120	78.43	0.040
	406.4		394	55,160	3.622	713.6	0.114	1.287	90.09	0.040
	457.2		445	62,300	4.621	1,028	0.128	1.454	101.8	0.040
	508		495	69,300	5.717	1,415	0.143	1.617	113.2	0.040
	558.8		546	76,440	6.956	1,899	0.158	1.784	124.9	0.040
	609.6		597	83,580	8.316	2,482	0.172	1.950	136.5	0.040
203.2	304.8	191	292	55,772	2.714	396.3	0.084	1.775	169.6	0.055
	355.6		343	65,513	3.745	642.3	0.099	2.085	199.2	0.055
	406.4		394	75,254	4.942	973.5	0.114	2.396	228.8	0.055
	457.2		445	84,995	6.304	1,403	0.128	2.706	258.4	0.055
	508		495	94,545	7.800	1,930	0.143	3.010	287.4	0.055
	558.8		546	104,286	9.490	2,591	0.158	3.320	317.0	0.055
	609.6		597	114,027	11.35	3,387	0.172	3.630	346.7	0.055
254	355.6	241	343	82,663	4.726	810.4	0.099	3.320	400.1	0.070
	406.4		394	94,954	6.235	1,228	0.114	3.814	459.6	0.070
	457.2		445	107,245	7.954	1,770	0.128	4.308	519.1	0.070
	508		495	119,295	9.842	2,436	0.143	4.792	577.4	0.070
	558.8		546	131,586	11.97	3,269	0.158	5.285	636.9	0.070
	609.6		597	143,877	14.32	4,273	0.172	5.779	696.4	0.070
304.8	406.4	292	394	115,048	7.555	1,488	0.114	5.599	817.5	0.084
	457.2		445	129,940	9.637	2,144	0.128	6.324	923.3	0.084
	508		495	144,540	11.92	2,951	0.143	7.034	1,027	0.084
	558.8		546	159,432	14.51	3,961	0.158	7.759	1,133	0.084
	609.6		597	174,324	17.35	5,178	0.172	8.484	1,239	0.084

Nominal Size		*Dressed Size*			*X–X Axis*				*Y–Y Axis*		
b	d	b	d	A	S_{xx} $\times 10^6$	I_{xx} $\times 10^6$	r_{xx}	S_{yy} $\times 10^6$	I_{yy} $\times 10^6$	r_{yy}
(mm)	*(mm)*	*(mm)*	*(mm)*	*(mm²)*	*(mm³)*	*(mm⁴)*	*(mm)*	*(mm³)*	*(mm⁴)*	*(mm)*
Beams										
355.6	457.2	343	445	152,635	11.32	2,519	0.128	8.726	1,496	0.099
	508		495	169,785	14.01	3,467	0.143	9.706	1,665	0.099
	558.8		546	187,278	17.04	4,653	0.158	10.706	1,836	0.099
	609.6		597	204,771	20.37	6,082	0.172	11.706	2,008	0.099
406.4	508	394	495	195,030	16.09	3,982	0.143	12.807	2,523	0.114
	558.8		546	215,124	19.58	5,344	0.158	14.126	2,783	0.114
	609.6		597	235,218	23.40	6,986	0.172	15.446	3,043	0.114
457.2	558.8	445	546	242,970	22.11	6,036	0.158	18.020	4,010	0.128
	609.6		597	265,665	26.43	7,890	0.172	19.703	4,384	0.128
508	609.6	495	597	295,515	29.40	8,777	0.172	24.380	6,034	0.143

Table A1.2 Glulam section properties of selected Western Species sizes

Imperial Units

b (in)	d (in)	A (in²)	X–X Axis			Y–Y Axis		
			S_{xx} (in³)	I_{xx} (in⁴)	r_{xx} (in)	S_{yy} (in³)	I_{yy} (in⁴)	r_{yy} (in)
3⅛	9	28.13	42.2	189.8	2.598	14.65	22.9	0.902
	12	37.50	75.0	450	3.464	19.53	30.5	0.902
	15	46.9	117.2	879	4.330	24.4	38.1	0.902
	18	56.3	168.8	1,519	5.196	29.3	45.8	0.902
	21	65.6	229.7	2,412	6.062	34.2	53.4	0.902
	24	75.0	300.0	3,600	6.928	39.1	61.0	0.902
5⅛	12	61.50	123.0	738.0	3.464	52.53	134.6	1.479
	15	76.88	192.2	1,441	4.330	65.66	168.3	1.479
	18	92.3	276.8	2,491	5.196	78.8	201.9	1.479
	21	107.6	376.7	3,955	6.062	91.9	235.6	1.479
	27	138.4	622.7	8,406	7.794	118.2	302.9	1.479
	33	169.1	930.2	15,348	9.526	144.5	370.2	1.479
6¾	18	121.5	364.5	3,281	5.196	136.7	461.3	1.949
	24	162.0	648.0	7,776	6.928	182.3	615.1	1.949
	30	202.5	1,013	15,188	8.660	227.8	768.9	1.949
	36	243.0	1,458	26,244	10.39	273.4	922.6	1.949
	42	283.5	1,985	41,675	12.12	318.9	1,076	1.949
	48	324.0	2,592	62,208	13.86	364.5	1,230	1.949
	54	364.5	3,281	88,574	15.59	410.1	1,384	1.949
	60	405.0	4,050	121,500	17.32	455.6	1,538	1.949

			X–X Axis			Y–Y Axis		
b (in)	d (in)	A (in²)	S_{xx} (in³)	I_{xx} (in⁴)	r_{xx} (in)	S_{yy} (in³)	I_{yy} (in⁴)	r_{yy} (in)
8¼	24	210.0	840.0	10,080	6.928	306.3	1,340	2.526
	30	262.5	1,313	19,688	8.660	382.8	1,675	2.526
	36	315.0	1,890	34,020	10.39	459.4	2,010	2.526
	42	367.5	2,573	54,023	12.12	535.9	2,345	2.526
	48	420.0	3,360	80,640	13.86	612.5	2,680	2.526
	54	472.5	4,253	114,818	15.59	689.1	3,015	2.526
	60	525.0	5,250	157,500	17.32	765.6	3,350	2.526
10¾	30	322.5	1,613	24,188	8.660	577.8	3,106	3.103
	36	387.0	2,322	41,796	10.39	693.4	3,727	3.103
	42	451.5	3,161	66,371	12.12	808.9	4,348	3.103
	48	516.0	4,128	99,072	13.86	924.5	4,969	3.103
	54	580.5	5,225	141,062	15.59	1,040	5,590	3.103
	60	645.0	6,450	193,500	17.32	1,156	6,211	3.103

Table A1.2 Glulam section properties of selected Western Species sizes

Metric Units

			X–X Axis			Y–Y Axis		
b	d	A	S_{xx} $\times 10^6$	I_{xx} $\times 10^6$	r_{xx}	S_{yy} $\times 10^6$	I_{yy} $\times 10^6$	r_{yy}
(mm)	(mm)	(mm²)	(mm³)	(mm⁴)	(mm)	(mm³)	(mm⁴)	(mm)
79	229	18,145	0.691	79.0	0.066	0.240	9.53	0.023
	305	24,194	1.229	187.3	0.088	0.320	12.70	0.023
	381	30,242	1.920	366	0.110	0.400	15.9	0.023
	457	36,290	2.765	632	0.132	0.480	19.1	0.023
	533	42,339	3.76	1,004	0.154	0.560	22.2	0.023
	610	48,387	4.92	1,498	0.176	0.640	25.4	0.023
130	305	39,677	2.016	307.2	0.088	0.861	56.03	0.038
	381	49,597	3.149	600.0	0.110	1.076	70.04	0.038
	457	59,516	4.535	1,037	0.132	1.291	84.0	0.038
	533	69,435	6.173	1,646	0.154	1.506	98.1	0.038
	686	89,274	10.20	3,499	0.198	1.937	126.1	0.038
	838	109,113	15.24	6,388	0.242	2.367	154.1	0.038
171	457	78,387	5.973	1,365	0.132	2.240	192.0	0.049
	610	104,516	10.62	3,237	0.176	2.987	256.0	0.049
	762	130,645	16.59	6,322	0.220	3.733	320.0	0.049
	914	156,774	23.89	10,924	0.264	4.480	384.0	0.049
	1,067	182,903	32.52	17,346	0.308	5.226	448.0	0.049
	1,219	209,032	42.48	25,893	0.352	5.973	512.0	0.049
	1,372	235,161	53.76	36,867	0.396	6.720	576.0	0.049
	1,524	261,290	66.37	50,572	0.440	7.466	640.1	0.049

Table A1.2 *continued*

b (mm)	d (mm)	A (mm²)	X–X Axis S_{xx} ×10⁶ (mm³)	X–X Axis I_{xx} ×10⁶ (mm⁴)	X–X Axis r_{xx} (mm)	Y–Y Axis S_{yy} ×10⁶ (mm³)	Y–Y Axis I_{yy} × 10⁶ (mm⁴)	Y–Y Axis r_{yy} (mm)
222	610	135,484	13.77	4,196	0.176	5.019	557.7	0.064
	762	169,355	21.51	8,195	0.220	6.273	697.1	0.064
	914	203,225	30.97	14,160	0.264	7.528	836.5	0.064
	1,067	237,096	42.16	22,486	0.308	8.782	975.9	0.064
	1,219	270,967	55.06	33,565	0.352	10.04	1,115	0.064
	1,372	304,838	69.69	47,791	0.396	11.29	1,255	0.064
	1,524	338,709	86.03	65,556	0.440	12.55	1,394	0.064
273	762	208,064	26.42	10,068	0.220	9.469	1,293	0.079
	914	249,677	38.05	17,397	0.264	11.36	1,551	0.079
	1,067	291,290	51.79	27,625	0.308	13.26	1,810	0.079
	1,219	332,903	67.65	41,237	0.352	15.15	2,068	0.079
	1,372	374,515	85.61	58,714	0.396	17.04	2,327	0.079
	1,524	416,128	105.7	80,541	0.440	18.94	2,585	0.079

Table A1.3 Structural composite lumber section properties

Imperial Units

b (in)	d (in)	A (in²)	X–X Axis			Y–Y Axis		
			S_{xx} (in³)	I_{xx} (in⁴)	r_{xx} (in)	S_{yy} (in³)	I_{yy} (in⁴)	r_{yy} (in)
$1\frac{3}{4}$	$5\frac{1}{2}$	9.625	8.823	24.26	1.588	2.807	2.456	0.505
	$7\frac{1}{4}$	12.69	15.33	55.57	2.093	3.701	3.238	0.505
	$9\frac{1}{2}$	16.63	26.32	125.0	2.742	4.849	4.243	0.505
	$11\frac{7}{8}$	20.78	41.13	244.2	3.428	6.061	5.304	0.505
	14	24.50	57.17	400.2	4.041	7.146	6.253	0.505
	16	28.00	74.67	597.3	4.619	8.167	7.146	0.505
	18	31.50	94.50	850.5	5.20	9.188	8.039	0.505
$3\frac{1}{2}$	$3\frac{1}{2}$	12.25	7.146	12.51	1.010	7.146	12.51	1.010
	$4\frac{3}{8}$	15.31	11.17	24.42	1.263	8.932	15.63	1.010
	$5\frac{1}{4}$	18.38	16.08	42.21	1.516	10.72	18.76	1.010
	7	24.50	28.58	100.0	2.021	14.29	25.01	1.010
	$8\frac{5}{8}$	30.19	43.39	187.1	2.490	17.61	30.82	1.010
	$9\frac{1}{2}$	33.25	52.65	250.1	2.742	19.40	33.94	1.010
	$11\frac{7}{8}$	41.56	82.26	488.4	3.428	24.24	42.43	1.010
	14	49.00	114.3	800.3	4.041	28.58	50.02	1.010
	16	56.00	149.3	1,195	4.619	32.67	57.17	1.010
	18	63.00	189.0	1,701	5.196	36.75	64.31	1.010

Table A1.3 *continued*

b (in)	d (in)	A (in²)	S_{xx} (in³)	I_{xx} (in⁴)	r_{xx} (in)	S_{yy} (in³)	I_{yy} (in⁴)	r_{yy} (in)
			X–X Axis			Y–Y Axis		
5¼	9½	49.88	78.97	375.1	2.742	43.64	114.6	1.516
	11⅞	62.34	123.4	732.6	3.428	54.55	143.2	1.516
	14	73.50	171.5	1,201	4.041	64.31	168.8	1.516
	16	84.00	224.0	1,792	4.62	73.50	192.9	1.516
	18	94.50	283.5	2,552	5.20	82.69	217.1	1.516
7	9½	66.50	105.3	500.1	2.742	77.58	271.5	2.021
	11⅞	83.13	164.5	976.8	3.428	96.98	339.4	2.021
	14	98.00	228.7	1,601	4.041	114.3	400.2	2.021
	16	112.0	298.7	2,389	4.62	130.7	457.3	2.021
	18	126.0	378.0	3,402	5.20	147.0	514.5	2.021

Appendix 1

Table A1.3 Structural composite lumber section properties

Metric Units

b	d	A	X–X Axis			Y–Y Axis		
			S_{xx} $\times 10^6$	I_{xx} $\times 10^6$	r_{xx}	S_{yy} $\times 10^6$	I_{yy} $\times 10^6$	r_{yy}
(mm)	(mm)	(mm^2)	(mm^2)	(mm^4)	(mm)	(mm^3)	(mm^4)	(mm)
44	140	6,210	0.145	10.10	40.33	0.046	1.022	12.83
	184	8,185	0.251	23.13	53.16	0.061	1.348	12.83
	241	10,726	0.431	52.04	69.66	0.079	1.766	12.83
	302	13,407	0.674	101.6	87.07	0.099	2.208	12.83
	356	15,806	0.937	166.6	102.7	0.117	2.603	12.83
	406	18,064	1.224	248.6	117.3	0.134	2.974	12.83
	457	20,323	1.549	354.0	132.0	0.151	3.346	12.83
89	89	7,903	0.117	5.205	25.66	0.117	5.205	25.66
	111	9,879	0.183	10.17	32.08	0.146	6.506	25.66
	133	11,855	0.263	17.57	38.49	0.176	7.808	25.66
	178	15,806	0.468	41.64	51.33	0.234	10.41	25.66
	219	19,476	0.711	77.89	63.24	0.289	12.83	25.66
	241	21,452	0.863	104.1	69.66	0.318	14.13	25.66
	302	26,814	1.348	203.3	87.07	0.397	17.66	25.66
	356	31,613	1.874	333.1	102.7	0.468	20.82	25.66
	406	36,129	2.447	497.3	117.3	0.535	23.79	25.66
	457	40,645	3.097	708.0	132.0	0.602	26.77	25.66

Table A1.3 *continued*

b	d	A	S_{xx} ×10⁶	I_{xx} ×10⁶	r_{xx}	S_{yy} ×10⁶	I_{yy} ×10⁶	r_{yy}
(mm)	(mm)	(mm²)	(mm²)	(mm⁴)	(mm)	(mm³)	(mm⁴)	(mm)
133	241	32,177	1.294	156.1	69.66	0.715	47.68	38.49
	302	40,222	2.022	304.9	87.07	0.894	59.60	38.49
	356	47,419	2.810	499.7	102.7	1.054	70.27	38.49
	406	54,193	3.671	745.9	117.3	1.204	80.31	38.49
	457	60,968	4.646	1,062	132.0	1.355	90.34	38.49
178	241	42,903	1.725	208.2	69.66	1.271	113.0	51.33
	302	53,629	2.696	406.6	87.07	1.589	141.3	51.33
	356	63,226	3.747	666.2	102.7	1.874	166.6	51.33
	406	72,258	4.894	994.5	117.3	2.141	190.4	51.33
	457	81,290	6.194	1,416	132.0	2.409	214.2	51.33

The header spans: *X–X Axis* covers S_{xx}, I_{xx}, r_{xx}; *Y–Y Axis* covers S_{yy}, I_{yy}, r_{yy}.

Timber Reference Design Values

Appendix 2

Table A2.1 Reference values for visually graded dimension lumber (2–4 in)

		Imperial Units							
Species & Grade	Size Class	Bending F_b (lb/in²)	Tension Parallel to Grain, F_t (lb/in²)	Shear, F_v (lb/in²)	Comp Perpendicular to Grain, $F_{c\perp}$ (lb/in²)	Comp Parallel to Grain, F_c (lb/in²)	Modulus of Elasticity E (lb/in²)	Modulus of Elasticity E_{min} (lb/in²)	Specific Gravity, G
Aspen									
Select structural	2 in & wider	875	500	120	265	725	1,100,000	400,000	0.39
No. 1		625	375	120	265	600	1,100,000	400,000	
No. 2		600	350	120	265	450	1,000,000	370,000	
No. 3		350	200	120	265	275	900,000	330,000	
Stud		475	275	120	265	300	900,000	330,000	
Construction	2–4 in	700	400	120	265	625	900,000	330,000	
Standard	wide	375	225	120	265	475	900,000	330,000	
Douglas Fir–Larch (North)									
Select structural	2 in & wider	1,350	825	180	625	1,900	1,900,000	690,000	0.49
No. 1 & better		1,150	750	180	625	1,800	1,800,000	660,000	
No. 1/no. 2		850	500	180	625	1,400	1,600,000	580,000	
No. 3		475	300	180	625	825	1,400,000	510,000	
Stud		650	400	180	625	900	1,400,000	510,000	
Construction	2–4 in	950	575	180	625	1,800	1,500,000	550,000	
Standard	wide	525	325	180	625	1,450	1,400,000	510,000	

Imperial Units

Species & Grade	Size Class	Bending F_b (lb/in²)	Tension Parallel to Grain, F_t (lb/in²)	Shear, F_v (lb/in²)	Comp Perpendicular to Grain, $F_{c\perp}$ (lb/in²)	Comp Parallel to Grain, F_c (lb/in²)	Modulus of Elasticity E (lb/in²)	Modulus of Elasticity E_{min} (lb/in²)	Specific Gravity, G
Redwood									
Clear structural	2 in & wider	1,750	1,000	160	650	1,850	1,400,000	510,000	0.44
Select structural		1,350	800	160	650	1,500	1,400,000	510,000	0.44
No. 1		975	575	160	650	1,200	1,300,000	470,000	0.44
No. 1 open grain		775	450	160	425	900	1,100,000	400,000	0.37
No. 3		525	300	160	650	550	1,100,000	400,000	0.44
Stud		575	325	160	425	450	900,000	330,000	0.44
Construction	2–4 in wide	825	475	160	425	925	900,000	330,000	0.44
Standard		450	275	160	425	725	900,000	330,000	0.44
Spruce–Pine–Fir									
Select structural	2 in & wider	1,250	700	135	425	1,400	1,500,000	550,000	0.42
No. 1/no. 2		875	450	135	425	1,150	1,400,000	510,000	
No. 3		500	250	135	425	650	1,200,000	440,000	
Stud		675	350	135	425	725	1,200,000	440,000	
Construction	2–4 in wide	1,000	500	135	425	1,400	1,300,000	470,000	
Standard		550	275	135	425	1,150	1,200,000	440,000	

Source: NDS 2015

Table A2.1m Reference values for visually graded dimension lumber (50–100mm)

	Metric Units										
Species & Grade	Size Class	Bending F_b (kN/m²)	Tension Parallel to Grain, F_t (kN/m²)	Shear, F_v (kN/m²)	Comp Perpendicular to Grain, $F_{c\perp}$ (kN/m²)	Comp Parallel to Grain, F_c (kN/m²)	Modulus of Elasticity E (kN/m²)	Modulus of Elasticity E_{min} (kN/m²)	Specific Gravity, G
Aspen									
Select structural	50 mm	6,033	3,447	827	1,827	4,999	7,584,232	2,757,903	0.39
No. 1	& wider	4,309	2,586	827	1,827	4,137	7,584,232	2,757,903	
No. 2		4,137	2,413	827	1,827	3,103	6,894,756	2,551,060	
No. 3		2,413	1,379	827	1,827	1,896	6,205,281	2,275,270	
Stud		3,275	1,896	827	1,827	2,068	6,205,281	2,275,270	
Construction	50–100-	4,826	2,758	827	1,827	4,309	6,205,281	2,275,270	
Standard	mm wide	2,586	1,551	827	1,827	3,275	6,205,281	2,275,270	
Douglas Fir-Larch (North)									
Select Structural	50 mm	9,308	5,688	1,241	4,309	13,100	13,100,037	4,757,382	0.49
No. 1 & better	& wider	7,929	5,171	1,241	4,309	12,411	12,410,561	4,550,539	
No. 1/no. 2		5,861	3,447	1,241	4,309	9,653	11,031,610	3,998,959	
No. 3		3,275	2,068	1,241	4,309	5,688	9,652,659	3,516,326	
Stud		4,482	2,758	1,241	4,309	6,205	9,652,659	3,516,326	
Construction	50–100	6,550	3,964	1,241	4,309	12,411	10,342,135	3,792,116	
Standard	mm wide	3,620	2,241	1,241	4,309	9,997	9,652,659	3,516,326	

	Metric Units							
Species & Grade	Size Class	Bending F_b (kN/m²)	Tension Parallel to Grain, F_t (kN/m²)	Shear, F_v (kN/m²)	Comp Perpendicular to Grain, $F_\perp$ (kN/m²)	Comp Parallel to Grain, F_c (kN/m²)	Modulus of Elasticity E (kN/m²)	Modulus of Elasticity E_{min} (kN/m²)	Specific Gravity, G
---	---	---	---	---	---	---	---	---	---
Redwood									
Clear structural	50 mm	12,066	6,895	1,103	4,482	12,755	9,652,659	3,516,326	0.44
Select structural	& wider	9,308	5,516	1,103	4,482	10,342	9,652,659	3,516,326	0.44
No. 1		6,722	3,964	1,103	4,482	8,274	8,963,183	3,240,535	0.44
No. 1 open grain		5,343	3,103	1,103	2,930	6,205	7,584,232	2,757,903	0.37
No. 3		3,620	2,068	1,103	4,482	3,792	7,584,232	2,757,903	0.44
Stud		3,964	2,241	1,103	2,930	3,103	6,205,281	2,275,270	0.44
Construction	50–100	5,688	3,275	1,103	2,930	6,378	6,205,281	2,275,270	0.44
Standard	mm wide	3,103	1,896	1,103	2,930	4,999	6,205,281	2,275,270	0.44
Spruce–Pine–Fir									
Select structural	50 mm	8,618	4,826	931	2,930	9,653	10,342,135	3,792,116	0.42
No. 1/no. 2	& wider	6,033	3,103	931	2,930	7,929	9,652,659	3,516,326	
No. 3		3,447	1,724	931	2,930	4,482	8,273,708	3,033,693	
Stud		4,654	2,413	931	2,930	4,999	8,273,708	3,033,693	
Construction	50–100	6,895	3,447	931	2,930	9,653	8,963,183	3,240,535	
Standard	mm wide	3,792	1,896	931	2,930	7,929	8,273,708	3,033,693	

Source: NDS 2015

Table A2.2 Reference values for visually graded timbers (5"× 5" and larger)

Imperial Units

Species & Grade	Bending F_b (lb/in²)	Tension Parallel to Grain F_t (lb/in²)	Shear F_v (lb/in²)	Comp Perpendicular to Grain $F_{c\perp}$ (lb/in²)	Comp Parallel to Grain F_c (lb/in²)	Modulus of Elasticity E (lb/in²)	E_{min} (lb/in²)	Specific Gravity G
Bald Cypress								
Select structural	1,150	750	200	615	1,050	1,300,000	470,000	0.43
No. 1	1,000	675	200	615	925	1,300,000	470,000	
No. 2	625	425	175	615	600	1,000,000	370,000	
Douglas Fir–Larch (North)								
Select structural	1,500	1,000	170	625	1,150	1,600,000	580,000	0.49
No. 1	1,200	825	170	625	1,000	1,600,000	580,000	
No. 2	725	475	170	625	700	1,300,000	470,000	
Redwood								
Clear structural	1,850	1,250	145	650	1,650	1,300,000	470,000	0.44
No. 1	1,200	800	145	650	1,050	1,300,000	470,000	
No. 2	1,000	525	145	650	900	1,100,000	400,000	
Spruce–Pine–Fir								
Select structural	1,000	675	125	335	700	1,200,000	440,000	0.36
No. 1	800	550	125	335	625	1,200,000	440,000	
No. 2	475	325	125	335	425	1,000,000	370,000	

Source: NDS 2015

Table A2.2 Reference values for visually graded timbers (125 × 125 mm and larger)

Metric Units

Species & Grade	Bending F_b (kN/m²)	Tension Parallel to Grain F_t (kN/m²)	Shear F_v (kN/m²)	Comp Perpendicular to Grain $F_{c\perp}$ (kN/m²)	Comp Parallel to Grain F_c (kN/m²)	Modulus of Elasticity E (kN/m²)	E_{min} (kN/m²)	Specific Gravity G
Bald Cypress								
Select structural	7,929	5,171	1,379	4,240	7,239	8,963,183	3,240,535	0.43
No. 1	6,895	4,654	1,379	4,240	6,378	8,963,183	3,240,535	
No. 2	4,309	2,930	1,207	4,240	4,137	6,894,756	2,551,060	
Douglas Fir–Larch (North)								
Select structural	10,342	6,895	1,172	4,309	7,929	11,031,610	3,998,959	0.49
No. 1	8,274	5,688	1,172	4,309	6,895	11,031,610	3,998,959	
No. 2	4,999	3,275	1,172	4,309	4,826	8,963,183	3,240,535	
Redwood								
Clear structural	12,755	8,618	1,000	4,482	11,376	8,963,183	3,240,535	0.44
No. 1	8,274	5,516	1,000	4,482	7,239	8,963,183	3,240,535	
No. 2	6,895	3,620	1,000	4,482	6,205	7,584,232	2,757,903	
Spruce–Pine–Fir								
Select structural	6,895	4,654	862	2,310	4,826	8,273,708	3,033,693	0.36
No. 1	5,516	3,792	862	2,310	4,309	8,273,708	3,033,693	
No. 2	3,275	2,241	862	2,310	2,930	6,894,756	2,551,060	

Source: NDS 2015

Table A2.3 Reference values for mechanically graded Douglas Fir-Larch (North) dimension lumber

Imperial Units

Grade	Size Class	Bending F_b (lb/in²)	Tension Parallel to Grain, F_t (lb/in²)	Shear F_v (lb/in²)	Comp Perpendicular to Grain $F_\perp$ (lb/in²)	Comp Parallel to Grain F_c (lb/in²)	Modulus of Elasticity E (lb/in²)	E_{min} (lb/in²)	Specific Gravity, G
M-5	2 in and wider	900	500	180	625	1,050	1,100,000	510,000	0.49
M-8		1,300	700	180	625	1,500	1,300,000	610,000	0.49
M-11		1,550	850	180	625	1,675	1,500,000	700,000	0.49
M-14		1,800	1,000	180	625	1,750	1,700,000	790,000	0.49
M-17		1,950	1,300	180	625	2,050	1,700,000	790,000	0.49
M-20		2,000	1,600	180	625	2,100	1,900,000	890,000	0.49
M-23		2,400	1,900	180	625	1,975	1,800,000	840,000	0.49
M-26		2,800	1,800	180	670	2,150	2,000,000	930,000	0.53

Notes: Size factors are already incorporated into this table

Source: NDS 2015

Table A2.3 Reference values for mechanically graded Douglas Fir–Larch (North) dimension lumber

Metric Units

Grade	Size Class	Bending F_b (kN/m²)	Tension Parallel to Grain, F_t (kN/m²)	Shear F_v (kN/m²)	Comp Perpendicular to Grain $F_{c\perp}$ (kN/m²)	Comp Parallel to Grain F_c (kN/m²)	Modulus of Elasticity E (kN/m²)	E_{min} (kN/m²)	Specific Gravity, G
M-5	50 mm & wider	6,205	3,447	1,241	4,309	7,239	7,584,232	3,516,326	0.49
M-8		8,963	4,826	1,241	4,309	10,342	8,963,183	4,205,801	0.49
M-11		10,687	5,861	1,241	4,309	11,549	10,342,135	4,826,329	0.49
M-14		12,411	6,895	1,241	4,309	12,066	11,721,086	5,446,858	0.49
M-17		13,445	8,963	1,241	4,309	14,134	11,721,086	5,446,858	0.49
M-20		13,790	11,032	1,241	4,309	14,479	13,100,037	6,136,333	0.49
M-23		16,547	13,100	1,241	4,309	13,617	12,410,561	5,791,595	0.49
M-26		19,305	12,411	1,241	4,619	14,824	13,789,513	6,412,123	0.53

Notes: Size factors are already incorporated into this table

Source: NDS 2015

Table A2.4 Reference values for visually graded Southern Pine

Imperial Units

Grade	Size Class	Bending F_b (lb/in²)	Tension Parallel to Grain F_t (lb/in²)	Shear F_v (lb/in²)	Comp Perpendicular to Grain $F_\perp$ (lb/in²)	Comp Parallel to Grain, F_c (lb/in²)	Modulus of Elasticity E (lb/in²)	E_{min} (lb/in²)	Specific Gravit G
Dense select structural	2–4 in wide	2,700	1,900	175	660	2,050	1,900,000	690,000	0.55
Select structural		2,350	1,650	175	565	1,900	1,800,000	660,000	
No. 1		1,500	1,000	175	565	1,650	1,600,000	580,000	
No. 2 Dense		1,200	750	175	660	1,500	1,600,000	580,000	
No. 3 and stud		650	400	175	565	850	1,300,000	470,000	
Construction	4 in wide	875	500	175	565	1,600	1,400,000	510,000	
Standard		475	275	175	565	1,300	1,200,000	440,000	
Dense select structural	8 in wide	2,200	1,550	175	660	1,850	1,900,000	690,000	0.55
Select structural		1,950	1,350	175	565	1,700	1,800,000	660,000	
No. 1		1,250	800	175	565	1,500	1,600,000	580,000	
No. 2 dense		975	600	175	660	1,400	1,600,000	580,000	
No. 3 and stud		525	325	175	565	775	1,300,000	470,000	

Note: Size factors are already incorporated into this table

Table A2.4 Reference values for visually graded Southern Pine (Metric Measures)

Metric Units

Grade	Size Class	Bending F_b (kN/m²)	Tension Parallel to Grain F_t (kN/m²)	Shear F_v (kN/m²)	Comp Perpendicular to Grain $F_\perp$ (kN/m²)	Comp Parallel to Grain, F_c (kN/m²)	Modulus of Elasticity E (kN/m²)	E_{min} (kN/m²)	Specific Gravity G
Dense select structural	50–100 mm wide	18,616	13,100	1,207	4,551	14,134	13,100,037	4,757,382	0.55
Select structural		16,203	11,376	1,207	3,896	13,100	12,410,561	4,550,539	
No. 1		10,342	6,895	1,207	3,896	11,376	11,031,610	3,998,959	
No. 2 dense		8,274	5,171	1,207	4,551	10,342	11,031,610	3,998,959	
No. 3 and stud		4,482	2,758	1,207	3,896	5,861	8,963,183	3,240,535	
Construction	200 mm wide	6,033	3,447	1,207	3,896	11,032	9,652,659	3,516,326	
Standard		3,275	1,896	1,207	3,896	8,963	8,273,708	3,033,693	
Dense select structural	200 mm wide	15,168	10,687	1,207	4,551	12,755	13,100,037	4,757,382	0.55
Select structural		13,445	9,308	1,207	3,896	11,721	12,410,561	4,550,539	
No. 1		8,618	5,516	1,207	3,896	10,342	11,031,610	3,998,959	
No. 2 dense		6,722	4,137	1,207	4,551	9,653	11,031,610	3,998,959	
No. 3 and stud		3,620	2,241	1,207	3,896	5,343	8,963,183	3,240,535	

Note: Size factors are already incorporated into this table

Source: NDS 2015

Table A2.5 Reference values for glued laminated timbers

Imperial Units

Combination	Species	Bending Tension on Tension Face F_b+ (lb/in²)	Bending Tension on Compression on Face F_b- (lb/in²)	Tension Parallel to Grain F_t (lb/in²)	Shear F_v (lb/in²)	Comp Perpendicular to Grain $F_\perp$ (lb/in²)	Comp Parallel to Grain F_c (lb/in²)	Modulus of Elasticity E (lb/in²)	E_{min} (lb/in²)	Specific Gravity G
16F-V3	DF/DF	1,600	1,250	975	265	560	1,500	1,500,000	790,000	0.50
16F-V6	DF/DF	1,600	1,600	560	265	560	1,600	1,600,000	850,000	0.50
16F-V2	SP/SP	1,600	1,400	1,000	300	650	1,300	1,500,000	790,000	0.55
16F-V5	SP/SP	1,600	1,600	1,000	300	650	1,550	1,600,000	850,000	0.55
24F-V4	DF/DF	2,400	1,850	1,100	265	650	1,650	1,800,000	950,000	0.50
24F-V8	DF/DF	2,400	2,400	1,100	265	650	1,650	1,800,000	950,000	0.50
24F-V3	SP/SP	2,400	2,000	1,150	300	740	1,650	1,800,000	950,000	0.55
24F-V8	SP/SP	2,400	2,400	1,150	300	740	1,650	1,800,000	950,000	0.55

Note: These values are for loads applied to the strong (x) axis

Source: NDS 2015

Appendix 2

Table A2.5 Reference values for glued laminated timbers

Metric Units

| Combination | Species | Bending | | Tension Parallel to Grain F_t (kN/m²) | Shear F_v (kN/m²) | Comp Perpendicular to Grain $F_{c\perp}$ (kN/m²) | Comp Parallel to Grain F_c (kN/m²) | Modulus of Elasticity | | Specific Gravity G |
		Tension on Tension Face F_b+ (kN/m²)	Tension on Compression Face F_b- (kN/m²)					E (kN/m²)	E_{min} (kN/m²)	
16F-V3	DF/DF	11,032	8,618	6,722	1,827	3,861	10,342	10,342,135	5,446,858	0.50
16F-V6	DF/DF	11,032	11,032	3,861	1,827	3,861	11,032	11,031,610	5,860,543	0.50
16F-V2	SP/SP	11,032	9,653	6,895	2,068	4,482	8,963	10,342,135	5,446,858	0.55
16F-V5	SP/SP	11,032	11,032	6,895	2,068	4,482	10,687	11,031,610	5,860,543	0.55
24F-V4	DF/DF	16,547	12,755	7,584	1,827	4,482	11,376	12,410,561	6,550,019	0.50
24F-V8	DF/DF	16,547	16,547	7,584	1,827	4,482	11,376	12,410,561	6,550,019	0.50
24F-V3	SP/SP	16,547	13,790	7,929	2,068	5,102	11,376	12,410,561	6,550,019	0.55
24F-V8	SP/SP	16,547	16,547	7,929	2,068	5,102	11,376	12,410,561	6,550,019	0.55

Note: These values are for loads applied to the strong (x) axis

Source: NDS 2015

Table A2.6 Reference values for structural composite lumber

Imperial Units

Grade	Size Class (in)	Bending F_b (lb/in²)	Tension Parallel to Grain F_t (lb/in²)	Shear F_v (lb/in²)	Comp Perpendicular to Grain $F_{c\perp}$ (lb/in²)	Comp Parallel to Grain F_c (lb/in²)	Modulus of Elasticity E (lb/in²)	E_{min} (lb/in²)
LVL								
1.6E WS	0.75	2,140	1,240	285	750	2,100	1,600,000	813,000
1.9E WS	to	2,600	1,555	285	750	2,510	1,900,000	966,000
2.0E-2900 F$_b$ WS	3.50	2,900	1,660	285	750	2,635	2,000,000	1,017,000
2.0E SP/EUC		2,750	1,805	285	880	2,635	2,000,000	1,017,000
2.6E SP/EUC		3,675	2,485	285	880	3,270	2,600,000	1,312,000
PSL								
1.8E DF	Up	2,500	1,755	230	600	2,500	1,800,000	915,000
2.2E DF	to	2,900	2,025	290	750	2,900	2,200,000	1,118,000
2.1E SP	11.0	3,100	2,160	320	825	3,100	2,100,000	1,067,000
LSL								
1.3E	1.25	1,700	1,075	425	710	1,835	1,300,000	661,000
1.55E	to	2,325	1,600	525	900	2,170	1,550,000	788,000
1.9E	5.50	3,075	2,150	625	1,090	2,505	1,900,000	966,000
Rim board								
0.6E OSB	1.125	700	–	395	660	–	600,000	305,000

Notes: (1.) These values are representative of LVL material. For design, verify local availability and use those specific values. (2.) These values are for material oriented in the strong axis

Source: ICC-ES Report ESR-1387

Table A2.6m Reference values for structural composite lumber

Metric Units

Grade	Size Class (mm)	Bending F_b (kN/m²)	Tension Parallel to Grain F_t (kN/m²)	Shear F_v (kN/m²)	Comp Perpendicular to Grain $F_{c\perp}$ (kN/m²)	Comp Parallel to Grain F_c (kN/m²)	E (kN/m²)	Modulus of Elasticity E_{min} (kN/m²)
LVL								
1.6E WS	19	14,755	8,549	1,965	5,171	14,479	11,031,610	5,605,437
1.9E WS	to	17,926	10,721	1,965	5,171	17,306	13,100,037	6,660,335
2.0E-2900 F_b WS	89	19,995	11,445	1,965	5,171	18,168	13,789,513	7,011,967
2.0E SP/EUC		18,961	12,445	1,965	6,067	18,168	13,789,513	7,011,967
2.6E SP/EUC		25,338	17,133	1,965	6,067	22,546	17,926,366	9,045,920
PSL								
1.8E DF	Up	17,237	12,100	1,586	4,137	17,237	12,410,561	6,308,702
2.2E DF	to	19,995	13,962	1,999	5,171	19,995	15,168,464	7,708,338
2.1E SP	280	21,374	14,893	2,206	5,688	21,374	14,478,988	7,356,705
LSL								
1.3E	32	11,721	7,412	2,930	4,895	12,652	8,963,183	4,557,434
1.55E	to	16,030	11,032	3,620	6,205	14,962	10,686,872	5,433,068
1.9E	140	21,201	14,824	4,309	7,515	17,271	13,100,037	6,660,335
Rim Board								
0.6E OSB	29	4,826	–	2,723	4,551	–	4,136,854	2,102,901

Notes: (1.) These values are representative of LVL material. For design, verify local availability and use those specific values. (2.) These values are for material oriented in the strong axis

Source: ICC-ES Report ESR-1387

Table A2.7 Reference values for selected I-joists

Imperial Units

Depth	Series	Moment, M (ft-lb)	Vertical Shear, V (lb)	Bend Stiffness, EI × 10⁶ (in²-lb)	Shear Stiffness, K × 10⁶ (in-lb/in)
$9^{1}/_{2}$	5000–1.8	2,460	1,475	160	5
$11^{7}/_{8}$		3,150	1,625	250	6
14		3,735	1,825	390	8
$9^{1}/_{2}$	6000–1.8	3,165	1,575	190	5
$11^{7}/_{8}$		4,060	1,675	320	6
14		4,815	1,925	470	8
16		5,495	2,175	635	9
14	60–2.0	6,235	1,675	430	7
16		7,440	1,925	635	8
18		8,520	2,175	860	9
14	90–2.0	11,390	2,350	940	8
16		13,050	2,550	1,275	9
18		14,690	2,750	1,660	10
20		16,310	2,850	2,100	11

Notes: (1.) These values are representative of I-joists. For design, verify local availability and use those specific values. (2.) Remember these values are allowable capacities, not allowable stresses. Compare them with moment and shear demands

Source: ICC-ES Report ESR-1336

Table A2.7 Reference values for selected I-joists

Metric Units

Depth	Series	Moment, M (N-m)	Vertical Shear, V (N)	Bend Stiffness, EI × 10⁶ (mm²-kN)	Shear Stiffness, K × 10⁶ (mm-lb/mm)
241	5000–1.8	3,335	6,561	459	22
302		4,271	7,228	717	27
356		5,064	8,118	1,119	36
241	6000–1.8	4,291	7,006	545	22
302		5,505	7,451	918	27
356		6,528	8,563	1,349	36
406		7,450	9,675	1,822	40
356	60–2.0	8,453	7,451	1,234	31
406		10,087	8,563	1,822	36
457		11,551	9,675	2,468	40
356	90–2.0	15,443	10,453	2,698	36
406		17,693	11,343	3,659	40
457		19,917	12,233	4,764	44
508		22,113	12,677	6,027	49

Notes: (1.) These values are representative of I-joists. For design, verify local availability and use those specific values. (2.) Remember these values are allowable capacities, not allowable stresses. Compare them with moment and shear demands

Source: ICC-ES Report ESR-1336

Connection Reference Design Values

Appendix 3

Table A3.1 Nail and spike reference withdrawal values *W*

$$W' = WC_DC_MC_tC_{eg}C_{tn}$$

Imperial Units

Specific Gravity	Withdrawal Strength (lb) Common Nail Size								
G	6d	8d	10d	16d	20d	30d	40d	50d	
	Diameter (in)								
	0.113 (2.87)	0.131 (3.33)	0.148 (3.76)	0.162 (4.11)	0.192 (4.88)	0.207 (5.26)	0.225 (5.72)	0.244 (6.20)	0.375 (9.53)
0.55	35	41	46	50	59	64	70	76	116
0.49	26	30	34	38	45	48	52	57	87
0.42	18	21	23	26	30	33	35	38	59
0.36	12	14	16	17	21	22	24	26	40

Note: Values are for each inch of penetration in side grain

Metric Units

Specific Gravity	Withdrawal Strength (N) Common Nail Size								
G	6d	8d	10d	16d	20d	30d	40d	50d	
	Diameter (mm)								
	2.87	3.33	3.76	4.11	4.88	5.26	5.72	6.20	9.53
0.55	61	72	81	88	103	112	123	133	203
0.49	46	53	60	67	79	84	91	100	152
0.42	32	37	40	46	53	58	61	67	103
0.36	21	25	28	30	37	39	42	46	70

Note: Values are for each 10 mm of thread penetration in side grain

Source: NDS 2015

Table A3.2 Lag screw reference withdrawal values, *W*

$$W' = WC_D C_M C_t C_{eg} C_{tn}$$

Imperial Units

Specific Gravity	Withdrawal Strength (lb) Lag Screw Diameter (in)					
G	¼	⁵⁄₁₆	³⁄₈	½	⁵⁄₈	¾
0.55	260	307	352	437	516	592
0.49	218	258	296	367	434	498
0.42	173	205	235	291	344	395
0.36	137	163	186	231	273	313

Note: (1.) Values are for each inch of thread penetration in side grain. (2.) Do not include the tapered tip in the length

Metric Units

Specific Gravity	Withdrawal Strength (N) Lag Screw Diameter (mm)					
G	6.4	7.9	9.5	12.7	15.9	19.1
0.55	455	538	616	765	904	1,037
0.49	382	452	518	643	760	872
0.42	303	359	412	510	602	692
0.36	240	285	326	405	478	548

Notes: (1.) Values are for each 10 mm of thread penetration in side grain. (2.) Do not include the tapered tip in the length

Source: NDS 2015

Table A3.3 Dowel bearing strength, F_e

$$Z' = Z C_D C_M C_t C_g C_\Delta C_{eg} C_{di} C_{tn}$$

Imperial Units

Specific Gravity			Dowel Bearing Strength (lb/in²)							
G	F_e	$F_{e\parallel}$	$F_{e\perp}$							
			Dowel Diameter (in)							
	$< \frac{1}{4}$	$\frac{1}{4}-1$	$\frac{1}{4}$	$\frac{3}{16}$	$\frac{3}{8}$	$\frac{7}{16}$	$\frac{1}{2}$	$\frac{5}{8}$	$\frac{3}{4}$
0.55	5,550	6,150	5,150	4,600	4,200	3,900	3,650	3,250	2,950
0.49	4,450	5,500	4,350	3,900	3,550	3,300	3,050	2,750	2,500
0.42	3,350	4,700	3,450	3,100	2,850	2,600	2,450	2,200	2,000
0.36	2,550	4,050	2,750	2,500	2,250	2,100	1,950	1,750	1,600

Metric Units

Specific Gravity			Dowel Bearing Strength (kN/m²)							
G	F_e	$F_{e\parallel}$	$F_{e\perp}$							
			Dowel Diameter (mm)							
	< 6.35	$6.35-25.0$	6.35	7.94	9.53	11.12	12.7	15.88	19.05
0.55	38,266	42,403	35,508	31,716	28,958	26,890	25,166	22,408	20,340
0.49	30,682	37,921	29,992	26,890	24,476	22,753	21,029	18,961	17,237
0.42	23,097	32,405	23,787	21,374	19,650	17,926	16,892	15,168	13,790
0.36	17,582	27,924	18,961	17,237	15,513	14,479	13,445	12,066	11,032

Source: NDS 2015

Table A3.4 Common, box, sinker nail single shear, all wood, reference lateral design values Z

$$Z' = ZC_D C_M C_t C_g C_\Delta C_{eg} C_{di} C_{tn}$$

Imperial Units *(lb)*

Side Member t_s, (in)	Nail Diameter D, (in)	Nail Type			Southern Pine, $G = 0.55$	Douglas Fir–Larch (N), $G = 0.49$	Spruce–Pine–Fir $G = 0.42$	Eastern Softwoods, $G = 0.36$
		Common	*Box*	*Sinker*				
³⁄₄	0.099		6d	7d	61	54	47	38
	0.113	6d	8d	8d	79	71	57	46
	0.120			10d	89	77	62	50
	0.128		10d		101	84	68	56
	0.131	8d			104	87	70	58
	0.135		16d	12d	108	91	74	61
	0.148	10d	20d	16d	121	102	83	69
	0.162	16d	40d		138	117	96	80
	0.177			20d	153	130	107	90
1¹⁄₂	0.113		8d	8d	79	71	61	54
	0.120			10d	89	80	69	60
	0.128		10d		101	91	79	69
	0.131	8d			106	95	82	72
	0.135		16d	12d	113	101	88	76
	0.148	10d	20d	16d	128	115	100	87
	0.162	16d	40d		154	138	120	104
	0.177			20d	178	159	138	121
	0.192	20d		30d	185	166	144	126
	0.207	30d		40d	203	182	158	131
	0.225	40d			224	201	172	138
	0.244	50d		60d	230	206	175	141

Source: NDS 2015

Table A3.4 Common, box, sinker nail single shear, all wood, reference lateral design values Z

$$Z' = ZC_D C_M C_t C_g C_\Delta C_{eg} C_{di} C_{tn}$$

Metric Units (N)

Side Member t_s, (mm)	Nail Diameter D, (mm)	Common	Box	Sinker	Southern Pine, $G = 0.55$	Douglas Fir–Larch $(N), G = 0.49$	Spruce–Pine–Fir $G = 0.42$	Eastern Softwoods, $G = 0.36$
19.1	2.51		6d	7d	271	240	209	169
	2.87	6d	8d	8d	351	316	254	205
	3.05			10d	396	343	276	222
	3.25		10d		449	374	302	249
	3.33	8d			463	387	311	258
	3.43		16d	12d	480	405	329	271
	3.76	10d	20d	16d	538	454	369	307
	4.11	16d	40d		614	520	427	356
	4.50			20d	681	578	476	400
38.1	2.87		8d	8d	351	316	271	240
	3.05			10d	396	356	307	267
	3.25		10d		449	405	351	307
	3.33	8d			472	423	365	320
	3.43		16d	12d	503	449	391	338
	3.76	10d	20d	16d	569	512	445	387
	4.11	16d	40d		685	614	534	463
	4.50			20d	792	707	614	538
	4.88	20d		30d	823	738	641	560
	5.26	30d		40d	903	810	703	583
	5.72	40d			996	894	765	614
	6.20	50d		60d	1,023	916	778	627

Source: NDS 2015

Table A3.5 Common, box, sinker nail single shear, steel side plate, reference lateral design values Z

$$Z' = Z C_D C_M C_t C_g C_\Delta C_{eg} C_{di} C_{tn}$$

Imperial Units *(lb)*

Side Member t_s, (in)	Nail Diameter D, (in)	Common Nail	Southern Pine, $G = 0.55$	Douglas Fir–Larch (N), $G = 0.49$	Spruce–Pine–Fir $G = 0.42$	Eastern Softwoods, $G = 0.36$
0.060	0.113	6d	79	72	63	56
(16 gage)	0.131	8d	104	95	83	73
	0.148	10d	126	114	100	88
	0.162	16d	150	135	119	105
	0.192	20d	179	162	142	125
0.120	0.113	6d	95	87	77	68
(11 gage)	0.131	8d	121	110	97	86
	0.148	10d	143	130	115	102
	0.162	16d	166	152	134	119
	0.192	20d	195	177	156	138
0.239	0.113	6d	107	97	84	74
(3 gage)	0.131	8d	144	130	114	99
	0.148	10d	174	157	137	121
	0.162	16d	209	188	165	145
	0.192	20d	251	227	198	174
	0.207	30d	270	246	217	191

Source: NDS 2015

Table A3.5 Common, box, sinker nail single shear, steel side plate, reference lateral design values Z

$$Z' = ZC_D C_M C_t C_g C_\Delta C_{eg} C_{di} C_{tn}$$

Metric Units (N)

Side Member t_s, (mm)	Nail Diameter D, (mm)	Common Nail	Southern Pine, $G = 0.55$	Douglas Fir–Larch (N), $G = 0.49$	Spruce–Pine–Fir $G = 0.42$	Eastern Softwoods, $G = 0.36$
1.52	2.87	6d	351	320	280	249
	3.33	8d	463	423	369	325
	3.76	10d	560	507	445	391
	4.11	16d	667	601	529	467
	4.88	20d	796	721	632	556
3.05	2.87	6d	423	387	343	302
	3.33	8d	538	489	431	383
	3.76	10d	636	578	512	454
	4.11	16d	738	676	596	529
	4.88	20d	867	787	694	614
6.07	2.87	6d	476	431	374	329
	3.33	8d	641	578	507	440
	3.76	10d	774	698	609	538
	4.11	16d	930	836	734	645
	4.88	20d	1,117	1,010	881	774
	5.26	30d	1,201	1,094	965	850

Source: NDS 2015

Table A3.6 Bolt, reference lateral design values Z in single shear all wood

Imperial Units

$$Z' = ZC_D C_M C_t C_g C_\Delta C_{eg} C_{di} C_{tn}$$

			Lateral Reference Strength (lb)															
Thickness			Southern Pine, G = 0.55				Douglas Fir–Larch (N), G = 0.49				Spruce–Pine–Fir, G = 0.42				Eastern Softwoods, G = 0.36			
Main Member t_m (in)	Side Member t_s (in)	Bolt Dia. D (in)	$Z_\parallel$	$Z_\perp$	$Z_{m\perp}$	$Z_\perp$	$Z_\parallel$	$Z_\perp$	$Z_{m\perp}$	$Z_\perp$	$Z_\parallel$	$Z_\perp$	$Z_{m\perp}$	$Z_\perp$	$Z_\parallel$	$Z_\perp$	$Z_{m\perp}$	$Z_\perp$
1½	1½	½	530	330	330	250	470	290	290	210	410	240	240	170	350	200	200	130
		⅝	660	400	400	280	590	350	350	240	510	290	290	190	440	240	240	150
		¾	800	460	460	310	710	400	400	260	610	340	340	210	520	280	280	170
		⅞	930	520	520	330	830	460	460	280	710	380	380	220	610	320	320	180
		1	1,060	580	580	350	950	510	510	300	810	430	430	240	700	360	360	190
3½	1½	½	660	400	470	360	610	360	420	320	540	320	370	280	490	280	300	250
		⅝	940	560	620	500	870	520	530	450	780	410	430	360	710	330	350	290
		¾	1,270	660	690	580	1,190	560	590	490	1,080	450	480	390	990	360	400	310
		⅞	1,680	720	770	630	1,570	600	650	530	1,340	490	540	420	1,160	390	440	340
		1	2,010	770	830	670	1,790	650	710	560	1,530	530	590	460	1,320	420	480	370

$$Z' = Z C_D C_M C_t C_g C_\Delta C_{eg} C_{di} C_{tn}$$

| Thickness | | | Lateral Reference Strength (lb) | | | | | | | | | | | | | | | |
Main Member t_m (in)	Side Member t_s (in)	Bolt Dia. D (in)	Southern Pine, G = 0.55				Douglas Fir–Larch (N), G = 0.49				Spruce–Pine–Fir, G = 0.42				Eastern Softwoods, G = 0.36			
			$Z_\parallel$	$Z_\perp$	$Z_{m\perp}$	$Z_{s\perp}$	$Z_\parallel$	$Z_\perp$	$Z_{m\perp}$	$Z_{s\perp}$	$Z_\parallel$	$Z_\perp$	$Z_{m\perp}$	$Z_{s\perp}$	$Z_\parallel$	$Z_{m\perp}$	$Z_\perp$	$Z_\perp$
5¼	1½	⅝	940	560	640	500	870	520	590	450	780	410	520	400	710	460	330	330
		¾	1,270	660	850	660	1,190	560	780	560	1,080	450	670	450	990	540	360	360
		⅞	1,680	720	1,060	720	1,570	600	900	600	1,440	490	730	490	1,330	600	390	390
		1	2,150	770	1,140	770	2,030	650	970	650	1,760	530	800	530	1,520	650	420	420
	3½	⅝	1,170	780	780	680	1,110	690	720	620	1,020	590	650	520	950	590	500	440
		¾	1,690	960	1,090	850	1,600	850	1,010	750	1,480	730	880	620	1,370	730	630	500
		⅞	2,300	1,160	1,380	1,000	2,170	1,040	1,190	840	1,920	910	990	670	1,710	830	800	550
		1	2,870	1,390	1,520	1,060	2,630	1,260	1,320	900	2,330	1,120	1,100	730	2,080	910	980	580
7½	1½	⅝	940	560	640	500	870	520	590	450	780	410	520	400	710	460	330	330
		¾	1,270	660	850	660	1,190	560	780	560	1,080	450	690	450	990	620	360	360
		⅞	1,680	720	1,090	720	1,570	600	990	600	1,440	490	890	490	1,330	800	390	390
		1	2,150	770	1,350	770	2,030	650	1,240	650	1,760	530	1,110	530	1,520	890	420	420
	3½	⅝	1,170	780	780	680	1,110	690	720	620	1,020	590	650	520	950	590	500	440
		¾	1,690	960	1,090	850	1,600	850	1,010	750	1,480	730	910	640	1,370	820	630	550
		⅞	2,300	1,160	1,450	1,020	2,170	1,040	1,340	840	1,920	910	1,180	780	1,710	980	800	680
		1	2,870	1,390	1,830	1,210	2,630	1,260	1,570	900	2,330	1,120	1,300	950	2,080	1,070	980	790

Source: NDS 2015

Appendix 3

Table A3.6m Bolt, reference lateral design values Z in single shear all wood

Metric Units

$$Z' = ZC_D C_M C_t C_g C_\Delta C_{eg} C_{di} C_{tn}$$

Lateral Reference Strength (N)

Thickness		Bolt Dia. D (mm)	Southern Pine, G = 0.55				Douglas Fir–Larch (N), G = 0.49				Spruce–Pine–Fir, G = 0.42				Eastern Softwoods, G = 0.36			
Main Member l_m (mm)	Side Member l_m (mm)		$Z_\parallel$	$Z_\perp$	$Z_{m\perp}$	$Z_\perp$	$Z_\parallel$	$Z_\perp$	$Z_{m\perp}$	$Z_\perp$	$Z_\parallel$	$Z_\perp$	$Z_{m\perp}$	$Z_\perp$	$Z_\parallel$	$Z_\perp$	$Z_{m\perp}$	$Z_\perp$
38.1	38.1	12.7	2,358	1,468	1,468	1,112	2,091	1,290	1,290	934	1,824	1,068	1,068	756	1,557	890	890	578
		15.9	2,936	1,779	1,779	1,246	2,624	1,557	1,557	1,068	2,269	1,290	1,290	845	1,957	1,068	1,068	667
		19.1	3,559	2,046	2,046	1,379	3,158	1,779	1,779	1,157	2,713	1,512	1,512	934	2,313	1,246	1,246	756
		22.2	4,137	2,313	2,313	1,468	3,692	2,046	2,046	1,246	3,158	1,690	1,690	979	2,713	1,423	1,423	801
		25.4	4,715	2,580	2,580	1,557	4,226	2,269	2,269	1,334	3,603	1,913	1,913	1,068	3,114	1,601	1,601	845
88.9	38.1	12.7	2,936	1,779	2,091	1,601	2,713	1,601	1,868	1,423	2,402	1,423	1,646	1,246	2,180	1,246	1,334	1,112
		15.9	4,181	2,491	2,758	2,224	3,870	2,313	2,358	2,002	3,470	1,824	1,913	1,601	3,158	1,468	1,557	1,290
		19.1	5,649	2,936	3,069	2,580	5,293	2,491	2,624	2,180	4,804	2,002	2,135	1,735	4,404	1,601	1,779	1,379
		22.2	7,473	3,203	3,425	2,802	6,984	2,669	2,891	2,358	5,961	2,180	2,402	1,868	5,160	1,735	1,957	1,512
		25.4	8,941	3,425	3,692	2,980	7,962	2,891	3,158	2,491	6,806	2,358	2,624	2,046	5,872	1,868	2,135	1,646

$$Z' = ZC_D C_M C_t C_g C_\Delta C_{eg} C_{di} C_{tn}$$

Lateral Reference Strength (N)

Thickness		Bolt Dia.	Southern Pine, G = 0.55				Douglas Fir–Larch (N), G = 0.49				Spruce–Pine–Fir, G = 0.42				Eastern Softwoods, G = 0.36			
Main Member t_m (mm)	Side Member t_s (mm)	D (mm)	$Z_\parallel$	$Z_\perp$	$Z_{m\perp}$	$Z_\perp$	$Z_\parallel$	$Z_\perp$	$Z_{m\perp}$	$Z_\perp$	$Z_\parallel$	$Z_\perp$	$Z_{m\perp}$	$Z_\perp$	$Z_\parallel$	$Z_\perp$	$Z_{m\perp}$	$Z_\perp$
133	38.1	15.9	4,181	2,491	2,847	2,224	3,870	2,313	2,624	2,002	3,470	1,824	2,313	1,779	3,158	1,468	2,046	1,468
		19.1	5,649	2,936	3,781	2,936	5,293	2,491	3,470	2,491	4,804	2,002	2,980	2,002	4,404	1,601	2,402	1,601
		22.2	7,473	3,203	4,715	3,203	6,984	2,669	4,003	2,669	6,405	2,180	3,247	2,180	5,916	1,735	2,669	1,735
		25.4	9,564	3,425	5,071	3,425	9,030	2,891	4,315	2,891	7,829	2,358	3,559	2,358	6,761	1,868	2,891	1,868
	38.1	15.9	5,204	3,470	3,470	3,025	4,938	3,069	3,203	2,758	4,537	2,624	2,891	2,313	4,226	2,224	2,624	1,957
		19.1	7,517	4,270	4,849	3,781	7,117	3,781	4,493	3,336	6,583	3,247	3,914	2,758	6,094	2,802	3,247	2,224
		22.2	10,231	5,160	6,139	4,448	9,653	4,626	5,293	3,737	8,541	4,048	4,404	2,980	7,606	3,559	3,692	2,447
		25.4	12,776	6,183	6,761	4,715	11,699	5,605	5,872	4,003	10,364	4,982	4,893	3,247	9,252	4,359	4,048	2,580
191	38.1	15.9	4,181	2,491	2,847	2,224	3,870	2,313	2,624	2,002	3,470	1,824	2,313	1,779	3,158	1,468	2,046	1,468
		19.1	5,649	2,936	3,781	2,936	5,293	2,491	3,470	2,491	4,804	2,002	3,069	2,002	4,404	1,601	2,758	1,601
		22.2	7,473	3,203	3,203	3,203	6,984	2,669	4,404	2,669	6,405	2,180	3,959	2,180	5,916	1,735	3,559	1,735
		25.4	9,564	3,425	6,005	3,425	9,030	2,891	5,516	2,891	7,829	2,358	4,938	2,358	6,761	1,868	3,959	1,868
	38.1	15.9	5,204	3,470	3,470	3,025	4,938	3,069	3,203	2,758	4,537	2,624	2,891	2,313	4,226	2,224	2,624	1,957
		19.1	7,517	4,270	4,849	3,781	7,117	3,781	4,493	3,336	6,583	3,247	4,048	2,847	6,094	2,802	3,648	2,447
		22.2	10,231	5,160	6,450	4,537	9,653	4,626	5,961	4,003	8,541	4,048	5,249	3,470	7,606	3,559	4,359	3,025
		25.4	12,776	6,183	8,140	5,382	11,699	5,605	6,984	4,804	10,364	4,982	5,783	4,226	9,252	4,359	4,760	3,514

Source: NDS 2015

Table A3.7 Bolt, reference lateral design values Z for single shear and steel plate side member

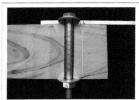

$$Z' = ZC_D C_M C_t C_g C_\Delta C_{eg} C_{di} C_{tn}$$

Imperial Units *(lb)*

Thickness		Bolt Dia.	Southern Pine, $G = 0.55$		Douglas Fir–Larch (N), $G = 0.49$		Spruce–Pine–Fir, $G = 0.42$		Eastern Softwoods, $G = 0.36$	
Main Member t_m (in)	Side Member t_s (in)	D (in)	$Z_\parallel$	$Z_\perp$	$Z_\parallel$	$Z_\perp$	$Z_\parallel$	$Z_\perp$	$Z_\parallel$	$Z_\perp$
$1\frac{1}{2}$	$\frac{1}{4}$	$\frac{1}{2}$	620	350	580	310	510	270	460	240
		$\frac{5}{8}$	780	400	720	360	640	320	580	280
		$\frac{3}{4}$	940	450	860	410	770	360	690	320
		$\frac{7}{8}$	1,090	510	1,010	450	900	400	810	360
		1	1,250	550	1,150	500	1,030	450	930	400
$3\frac{1}{2}$	$\frac{1}{4}$	$\frac{1}{2}$	860	550	820	510	770	430	720	360
		$\frac{5}{8}$	1,260	690	1,200	600	1,120	490	1,050	410
		$\frac{3}{4}$	1,740	760	1,660	660	1,450	540	1,260	450
		$\frac{7}{8}$	2,170	840	1,950	710	1,690	590	1,480	500
		1	2,480	890	2,230	770	1,930	650	1,690	540
$5\frac{1}{4}$	$\frac{1}{4}$	$\frac{5}{8}$	1,260	760	1,200	700	1,120	630	1,050	560
		$\frac{3}{4}$	1,740	1,000	1,660	930	1,550	760	1,450	620
		$\frac{7}{8}$	2,320	1,190	2,200	1,010	2,050	820	1,920	680
		1	2,980	1,270	2,840	1,080	2,640	890	2,450	730
$7\frac{1}{2}$	$\frac{1}{4}$	$\frac{5}{8}$	1,260	760	1,200	700	1,120	630	1,050	570
		$\frac{3}{4}$	1,740	1,000	1,660	930	1,550	840	1,450	750
		$\frac{7}{8}$	2,320	1,280	2,200	1,180	2,050	1,070	1,920	930
		1	2,980	1,590	2,840	1,470	2,640	1,230	2,470	1,000
$9\frac{1}{2}$	$\frac{1}{4}$	$\frac{3}{4}$	1,740	1,000	1,660	930	1,550	840	1,450	750
		$\frac{7}{8}$	2,320	1,280	2,200	1,180	2,050	1,070	1,920	970
		1	2,980	1,590	2,840	1,470	2,640	1,330	2,470	1,200

Source: NDS 2015

Table A3.7 Bolt, reference lateral design values Z for single shear and steel plate side member

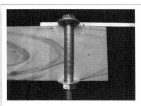

$$Z' = Z C_D C_M C_t C_g C_\Delta C_{eg} C_{di} C_{tn}$$

Metric Units *(N)*

Thickness		Bolt Dia.	Southern Pine, G = 0.55		Douglas Fir–Larch (N), G = 0.49		Spruce–Pine– Fir, G = 0.42		Eastern Softwoods, G = 0.36	
Main Member t_m*(mm)*	*Side Member* t_s *(mm)*	*D* *(mm)*	$Z_\parallel$	$Z_\perp$	$Z_\parallel$	$Z_\perp$	$Z_\parallel$	$Z_\perp$	$Z_\parallel$	$Z_\perp$
38.1	6.35	12.70	2,758	1,557	2,580	1,379	2,269	1,201	2,046	1,068
		15.88	3,470	1,779	3,203	1,601	2,847	1,423	2,580	1,246
		19.05	4,181	2,002	3,825	1,824	3,425	1,601	3,069	1,423
		22.23	4,849	2,269	4,493	2,002	4,003	1,779	3,603	1,601
		25.40	5,560	2,447	5,115	2,224	4,582	2,002	4,137	1,779
88.9	6.35	12.70	3,825	2,447	3,648	2,269	3,425	1,913	3,203	1,601
		15.88	5,605	3,069	5,338	2,669	4,982	2,180	4,671	1,824
		19.05	7,740	3,381	7,384	2,936	6,450	2,402	5,605	2,002
		22.23	9,653	3,737	8,674	3,158	7,517	2,624	6,583	2,224
		25.40	11,032	3,959	9,920	3,425	8,585	2,891	7,517	2,402
133.4	6.35	15.88	5,605	3,381	5,338	3,114	4,982	2,802	4,671	2,491
		19.05	7,740	4,448	7,384	4,137	6,895	3,381	6,450	2,758
		22.23	10,320	5,293	9,786	4,493	9,119	3,648	8,541	3,025
		25.40	13,256	5,649	12,633	4,804	11,743	3,959	10,898	3,247
190.5	6.35	15.88	5,605	3,381	5,338	3,114	4,982	2,802	4,671	2,535
		19.05	7,740	4,448	7,384	4,137	6,895	3,737	6,450	3,336
		22.23	10,320	5,694	9,786	5,249	9,119	4,760	8,541	4,137
		25.40	13,256	7,073	12,633	6,539	11,743	5,471	10,987	4,448
241.3	6.35	19.05	7,740	4,448	7,384	4,137	6,895	3,737	6,450	3,336
		22.23	10,320	5,694	9,786	5,249	9,119	4,760	8,541	4,315
		25.40	13,256	7,073	12,633	6,539	11,743	5,916	10,987	5,338

Source: NDS 2015

Table A3.8 Bolt, reference lateral design values Z in double shear, all wood

$$Z' = ZC_D C_M C_t C_g C_\Delta C_{eg} C_{di} C_{tn}$$

Imperial Units (lb)

Thickness		Bolt Dia.	Southern Pine, G = 0.55			Douglas Fir–Larch (N), G = 0.49			Spruce-Pine-Fir, G = 0.42			Eastern Softwoods, G = 0.36		
Main Member t_m (in)	Side Member t_s (in)	D (in)	$Z_\parallel$	Z	Z_m	$Z_\parallel$	$Z^\wedge$	Z_m	$Z_\parallel$	Z	Z_m	$Z_\parallel$	Z	Z_m
1½	1½	½	1,150	800	550	1,030	720	460	880	640	370	760	560	290
		⅝	1,440	1,130	610	1,290	1,030	520	1,100	830	410	950	660	330
		¾	1,730	1,330	660	1,550	1,130	560	1,320	900	450	1,140	720	360
		⅞	2,020	1,440	720	1,800	1,210	600	1,540	970	490	1,330	790	390
		1	2,310	1,530	770	2,060	1,290	650	1,760	1,050	530	1,520	840	420
3½	1½	½	1,320	800	940	1,210	720	850	1,080	640	740	980	560	660
		⅝	1,870	1,130	1,290	1,740	1,030	1,170	1,570	830	960	1,430	660	770
		¾	2,550	1,330	1,550	2,380	1,130	1,310	2,160	900	1,050	1,990	720	840
		⅞	3,360	1,440	1,680	3,150	1,210	1,410	2,880	970	1,130	2,660	790	920
		1	4,310	1,530	1,790	4,050	1,290	1,510	3,530	1,050	1,230	3,040	840	980

$Z' = ZC_DC_MC_tC_gC_\Delta C_{eg}C_{di}C_{tn}$														
Thickness		*Bolt Dia.*	*Southern Pine, G = 0.55*			*Douglas Fir–Larch (N), G = 0.49*			*Spruce–Pine–Fir, G = 0.42*			*Eastern Softwoods, G = 0.36)*		
Main Member t_m *(in)*	*Side Member* t_s *(in)*	*D (in)*	$Z_\parallel$	Z_s	Z_m	$Z_\parallel$	Z_s	Z_m	$Z_\parallel$	Z_s	Z_m	$Z_\parallel$	Z_s	Z_m
5¼	1½	⅝	1,870	1,130	1,290	1,740	1,030	1,170	1,570	830	1,040	1,430	660	920
		¾	2,550	1,330	1,690	2,380	1,130	1,550	2,160	900	1,380	1,990	720	1,230
		⅝	3,360	1,440	2,170	3,150	1,210	1,990	2,880	970	1,700	2,660	790	1,380
		1	4,310	1,530	2,680	4,050	1,290	2,260	3,530	1,050	1,840	3,040	840	1,470
	3½	⅝	2,340	1,560	1,560	2,220	1,390	1,450	2,050	1,170	1,310	1,900	1,000	1,150
		¾	3,380	1,910	2,180	3,190	1,700	1,970	2,950	1,460	1,580	2,740	1,270	1,260
		⅝	4,600	2,330	2,530	4,350	2,070	2,110	3,840	1,810	1,700	3,410	1,610	1,380
		1	5,740	2,780	2,680	5,250	2,520	2,260	4,660	2,240	1,840	4,170	1,960	1,470
7½	1½	⅝8	1,870	1,130	1,290	1,740	1,030	1,170	1,570	830	1,040	1,430	660	920
		¾	2,550	1,330	1,690	2,380	1,130	1,550	2,160	900	1,380	1,990	720	1,230
		⅝	3,360	1,440	2,170	3,150	1,210	1,990	2,880	970	1,780	2,660	790	1,600
		1	4,310	1,530	2,700	4,050	1,290	2,480	3,530	1,050	2,240	3,040	840	2,010
	3½	⅝	2,340	1,560	1,560	2,220	1,390	1,450	2,050	1,170	1,310	1,900	1,000	1,180
		¾	3,380	1,910	2,180	3,190	1,700	2,020	2,950	1,460	1,820	2,740	1,270	1,650
		⅞	4,600	2,330	2,890	4,350	2,070	2,670	3,840	1,810	2,420	3,410	1,610	1,970
		1	5,740	2,780	3,680	5,250	2,520	3,230	4,660	2,240	2,630	4,170	1,960	2,100

Source: *NDS* 2015

Table A3.8m Bolt, reference lateral design values Z in double shear, all wood

$$Z' = ZC_D C_M C_t C_g C_\Delta C_{eg} C_{di} C_{tn}$$

Metric Units (N)

Thickness		Bolt Dia.	Southern Pine, G = 0.55			Douglas Fir–Larch (N), G = 0.49			Spruce–Pine–Fir, G = 0.42			Eastern Softwoods, G = 0.36		
Main Member t_m (mm)	Side Member t_s (mm)	D (mm)	$Z_\parallel$	$Z_{s\perp}$	$Z_{m\perp}$	$Z_\parallel$	$Z_{s\perp}$	$Z_{m\perp}$	$Z_\parallel$	$Z_{s\perp}$	$Z_{m\perp}$	$Z_\parallel$	$Z_{s\perp}$	$Z_{m\perp}$
38.1	38.1	12.7	5,115	3,559	2,447	4,582	3,203	2,046	3,914	2,847	1,646	3,381	2,491	1,290
		15.9	6,405	5,026	2,713	5,738	4,582	2,313	4,893	3,692	1,824	4,226	2,936	1,468
		19.1	7,695	5,916	2,936	6,895	5,026	2,491	5,872	4,003	2,002	5,071	3,203	1,601
		22.2	8,985	6,405	3,203	8,007	5,382	2,669	6,850	4,315	2,180	5,916	3,514	1,735
		25.4	10,275	6,806	3,425	9,163	5,738	2,891	7,829	4,671	2,358	6,761	3,737	1,868
88.9	38.1	12.7	5,872	3,559	4,181	5,382	3,203	3,781	4,804	2,847	3,292	4,359	2,491	2,936
		15.9	8,318	5,026	5,738	7,740	4,582	5,204	6,984	3,692	4,270	6,361	2,936	3,425
		19.1	11,343	5,916	6,895	10,587	5,026	5,827	9,608	4,003	4,671	8,852	3,203	3,737
		22.2	14,946	6,405	7,473	14,012	5,382	6,272	12,811	4,315	5,026	11,832	3,514	4,092
		25.4	19,172	6,806	7,962	18,015	5,738	6,717	15,702	4,671	5,471	13,523	3,737	4,359

$$Z' = ZC_D C_M C_t C_g C_\Delta C_{eg} C_{di} C_{tn}$$

Thickness		Bolt Dia.	Southern Pine, G = 0.55			Douglas Fir–Larch (N), G = 0.49			Spruce–Pine–Fir, G = 0.42			Eastern Softwoods, G = 0.36		
Main Member t_m (in)	Side Member t_s (in)	D (in)	$Z_\parallel$	$Z_\perp$	Z_m	$Z_\parallel$	$Z_\perp$	Z_m	$Z_\parallel$	$Z_\perp$	Z_m	$Z_\parallel$	$Z_\perp$	Z_m
133	38.1	15.9	8,318	5,026	5,738	7,740	4,582	5,204	6,984	3,692	4,626	6,361	2,936	4,092
		19.1	11,343	5,916	7,517	10,587	5,026	6,895	9,608	4,003	6,139	8,852	3,203	5,471
		22.2	14,946	6,405	9,653	14,012	5,382	8,852	12,811	4,315	7,562	11,832	3,514	6,139
		25.4	19,172	6,806	11,921	18,015	5,738	10,053	15,702	4,671	8,185	13,523	3,737	6,539
	38.1	15.9	10,409	6,939	6,939	9,875	6,183	6,450	9,119	5,204	5,827	8,452	4,448	5,115
		19.1	15,035	8,496	9,697	14,190	7,562	8,763	13,122	6,494	7,028	12,188	5,649	5,605
		22.2	20,462	10,364	11,254	19,350	9,208	9,386	17,081	8,051	7,562	15,168	7,162	6,139
		25.4	25,533	12,366	11,921	23,353	11,210	10,053	20,729	9,964	8,185	18,549	8,719	6,539
191	38.1	15.9	8,318	5,026	5,738	7,740	4,582	5,204	6,984	3,692	4,626	6,361	2,936	4,092
		19.1	11,343	5,916	7,517	10,587	5,026	6,895	9,608	4,003	6,139	8,852	3,203	5,471
		22.2	14,946	6,405	9,653	14,012	5,382	8,852	12,811	4,315	7,918	11,832	3,514	7,117
		25.4	19,172	6,806	12,010	18,015	5,738	11,032	15,702	4,671	9,964	13,523	3,737	8,941
	38.1	15.9	10,409	6,939	6,939	9,875	6,183	6,450	9,119	5,204	5,827	8,452	4,448	5,249
		19.1	15,035	8,496	9,697	14,190	7,562	8,985	13,122	6,494	8,096	12,188	5,649	7,340
		22.2	20,462	10,364	12,855	19,350	9,208	11,877	17,081	8,051	10,765	15,168	7,162	8,763
		25.4	25,533	12,366	16,369	23,353	11,210	14,368	20,729	9,964	11,699	18,549	8,719	9,341

Source: NDS 2015

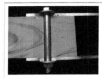

$$Z' = ZC_D C_M C_t C_g C_\Delta C_{eg} C_{di} C_{tn}$$

Imperial Units *(lb)*

Thickness		Bolt Dia.	Southern Pine, $G = 0.55$		Douglas Fir–Larch (N), $G = 0.49$		Spruce–Pine–Fir, $G = 0.42$		Eastern Softwoods, $G = 0.36$	
Main Member t_m (in)	Side Member t_s (in)	D (in)	$Z_\parallel$	$Z_\perp$	$Z_\parallel$	$Z_\perp$	$Z_\parallel$	$Z_\perp$	$Z_\parallel$	$Z_\perp$
$1\frac{1}{2}$	$\frac{1}{4}$	$\frac{1}{2}$	1,150	550	1,030	460	880	370	760	290
		$\frac{5}{8}$	1,440	610	1,290	520	1,100	410	950	330
		$\frac{3}{4}$	1,730	660	1,550	560	1,320	450	1,140	360
		$\frac{7}{8}$	2,020	720	1,800	600	1,540	490	1,330	390
		1	2,310	770	2,060	650	1,760	530	1,520	420
$3\frac{1}{2}$	$\frac{1}{4}$	$\frac{1}{2}$	1,720	1,100	1,640	1,010	1,530	860	1,430	680
		$\frac{5}{8}$	2,510	1,420	2,390	1,200	2,230	960	2,090	770
		$\frac{3}{4}$	3,480	1,550	3,320	1,310	3,080	1,050	2,660	840
		$\frac{7}{8}$	4,630	1,680	4,210	1,410	3,600	1,130	3,100	920
		1	5,380	1,790	4,810	1,510	4,110	1,230	3,540	980
$5\frac{1}{4}$	$\frac{1}{4}$	$\frac{5}{8}$	2,510	1,510	2,390	1,400	2,230	1,270	2,090	1,140
		$\frac{3}{4}$	3,480	2,000	3,320	1,850	3,090	1,580	2,890	1,260
		$\frac{7}{8}$	4,630	2,530	4,410	2,110	4,110	1,700	3,840	1,380
		1	5,960	2,680	5,670	2,260	5,280	1,840	4,930	1,470
$7\frac{1}{2}$	$\frac{1}{4}$	$\frac{5}{8}$	2,510	1,510	2,390	1,400	2,230	1,270	2,090	1,140
		$\frac{3}{4}$	3,480	2,000	3,320	1,850	3,090	1,670	2,890	1,500
		$\frac{7}{8}$	4,630	2,570	4,410	2,360	4,110	2,130	3,840	1,930
		1	5,960	3,180	5,670	2,940	5,280	2,630	4,930	2,100
$9\frac{1}{2}$	$\frac{1}{4}$	$\frac{3}{4}$	3,480	2,000	3,320	1,850	3,090	1,670	2,890	1,500
		$\frac{7}{8}$	4,630	2,570	4,410	2,360	4,110	2,130	3,840	1,930
		1	5,960	3,180	5,670	2,940	5,280	2,660	4,930	2,400
$11\frac{1}{2}$	$\frac{1}{4}$	$\frac{7}{8}$	4,630	2,570	4,410	2,360	4,110	2,130	3,840	1,930
		1	5,960	3,180	5,670	2,940	5,280	2,660	4,930	2,400
$13\frac{1}{2}$	$\frac{1}{4}$	1	5,960	3,180	5,670	2,940	5,280	2,660	4,930	2,400

Source: NDS 2015

Table A3.9 Bolt, double shear, and steel plate side member reference lateral design values Z

$$Z' = ZC_D C_M C_t C_g C_\Delta C_{eg} C_{di} C_{tn}$$

Metric Units (N)

Thickness		Bolt Dia.	Southern Pine, $G = 0.55$		Douglas Fir–Larch (N), $G = 0.49$		Spruce–Pine–Fir, $G = 0.42$		Eastern Softwoods, $G = 0.36$	
Main Member t_m(mm)	Side Member t_s(mm)	D (mm)	$Z_\parallel$	$Z_\perp$	$Z_\parallel$	$Z_\perp$	$Z_\parallel$	$Z_\perp$	$Z_\parallel$	$Z_\perp$
38.1	6.4	12.7	5,115	2,447	4,582	2,046	3,914	1,646	3,381	1,290
		15.9	6,405	2,713	5,738	2,313	4,893	1,824	4,226	1,468
		19.1	7,695	2,936	6,895	2,491	5,872	2,002	5,071	1,601
		22.2	8,985	3,203	8,007	2,669	6,850	2,180	5,916	1,735
		25.4	10,275	3,425	9,163	2,891	7,829	2,358	6,761	1,868
88.9	6.4	12.7	7,651	4,893	7,295	4,493	6,806	3,825	6,361	3,025
		15.9	11,165	6,316	10,631	5,338	9,920	4,270	9,297	3,425
		19.1	15,480	6,895	14,768	5,827	13,701	4,671	11,832	3,737
		22.2	20,595	7,473	18,727	6,272	16,014	5,026	13,789	4,092
		25.4	23,931	7,962	21,396	6,717	18,282	5,471	15,747	4,359
133.4	6.4	15.9	11,165	6,717	10,631	6,228	9,920	5,649	9,297	5,071
		19.1	15,480	8,896	14,768	8,229	13,745	7,028	12,855	5,605
		22.2	20,595	11,254	19,617	9,386	18,282	7,562	17,081	6,139
		25.4	26,511	11,921	25,221	10,053	23,487	8,185	21,930	6,539
190.5	6.4	15.9	11,165	6,717	10,631	6,228	9,920	5,649	9,297	5,071
		19.1	15,480	8,896	14,768	8,229	13,745	7,429	12,855	6,672
		22.2	20,595	11,432	19,617	10,498	18,282	9,475	17,081	8,585
		25.4	26,511	14,145	25,221	13,078	23,487	11,699	21,930	9,341
241.3	6.4	19.1	15,480	8,896	14,768	8,229	13,745	7,429	12,855	6,672
		22.2	20,595	11,432	19,617	10,498	18,282	9,475	17,081	8,585
		25.4	26,511	14,145	25,221	13,078	23,487	11,832	21,930	10,676
292.1	6.4	22.2	20,595	11,432	19,617	10,498	18,282	9,475	17,081	8,585
		25.4	26,511	14,145	25,221	13,078	23,487	11,832	21,930	10,676
342.9	6.4	25.4	26,511	14,145	25,221	13,078	23,487	11,832	21,930	10,676

Source: NDS 2015

Table A3.10 Lag screw, single shear, all wood, reference lateral design values Z

$$Z' = ZC_D C_M C_t C_g C_\Delta C_{eg} C_{di} C_{tn}$$

Imperial Units (lb)

Side Member t_s (in)	Bolt Dia. D (in)	Southern Pine, G = 0.55				Douglas Fir–Larch (N), G = 0.49				Spruce–Pine–Fir, G = 0.42				Eastern Softwoods, G = 0.36			
		$Z_\parallel$	$Z_{s\perp}$	$Z_{m\perp}$	$Z_\perp$	$Z_\parallel$	$Z_{s\perp}$	$Z_{m\perp}$	$Z_\perp$	$Z_\parallel$	$Z_{s\perp}$	$Z_{m\perp}$	$Z_\perp$	$Z_\parallel$	$Z_{s\perp}$	$Z_{m\perp}$	$Z_\perp$
3/4	1/4	150	110	120	110	140	100	110	90	120	80	90	80	110	70	80	70
	5/16	180	120	130	120	160	110	120	100	150	100	110	90	130	90	90	80
	3/8	180	120	130	110	170	110	120	100	150	90	110	90	130	80	90	70
1½	1/4	160	120	120	120	150	110	110	110	140	100	100	100	130	90	90	90
	5/16	210	150	150	140	200	140	140	130	180	130	130	120	170	110	120	100
	3/8	210	150	150	140	200	140	140	130	180	130	130	110	170	110	120	100
	1/2	410	250	290	230	390	220	260	200	350	190	240	170	310	160	210	150
	5/8	600	340	420	310	550	310	380	270	490	270	330	240	440	240	290	210
	3/4	830	470	560	410	760	430	510	370	690	350	440	330	620	280	390	280
3½	1/4	160	120	120	120	150	110	110	110	140	100	100	100	130	90	90	90
	5/16	210	150	150	140	200	140	140	130	180	130	130	120	170	120	120	110
	3/8	210	150	150	140	200	140	140	130	180	130	130	130	170	120	120	100
	1/2	410	290	290	250	390	260	260	230	360	240	240	230	340	220	220	190
	5/8	670	440	440	390	630	410	410	360	580	370	370	360	540	330	340	280
	3/4	1,010	650	650	560	950	580	600	510	880	490	540	510	820	420	490	370

Source: NDS 2015

Table A3.10 Lag screw, single shear, all wood, reference lateral design values Z

$$Z' = ZC_D C_M C_t C_g C_\Delta C_{eg} C_{di} C_{tn}$$

Metric Units (N)

Side Member t_s (mm)	Bolt Dia. D (mm)	Southern Pine, G = 0.55				Douglas Fir-Larch (N), G = 0.49				Spruce-Pine-Fir, G = 0.42				Eastern Softwoods, G = 0.36			
		$Z_\parallel$	$Z_\perp$	$Z_{m\perp}$	$Z_\perp$	$Z_\parallel$	$Z_\perp$	$Z_{m\perp}$	$Z_\perp$	$Z_\parallel$	$Z_{s\perp}$	$Z_{m\perp}$	$Z_\perp$	$Z_\parallel$	$Z_{s\perp}$	$Z_{m\perp}$	$Z_\perp$
19.1	6.35	667	489	534	489	623	445	489	400	534	356	400	400	489	311	356	311
	7.94	801	534	578	534	712	489	534	445	667	445	489	445	578	400	400	356
	9.53	801	534	578	489	756	489	534	445	667	400	489	445	578	356	400	311
38.1	6.35	712	534	534	534	667	489	489	489	623	445	445	489	578	400	400	400
	7.94	934	667	667	623	890	623	623	578	801	578	578	578	756	489	534	445
	9.53	934	667	667	623	890	623	623	578	801	578	578	578	756	489	534	445
	12.70	1,824	1,112	1,290	1,023	1,735	979	1,157	890	1,557	845	1,068	890	1,379	712	934	667
	15.88	2,669	1,512	1,868	1,379	2,447	1,379	1,690	1,201	2,180	1,201	1,468	1,201	1,957	1,068	1,290	934
	19.05	3,692	2,091	2,491	1,824	3,381	1,913	2,269	1,646	3,069	1,557	1,957	1,646	2,758	1,246	1,735	1,246
88.9	6.35	712	534	534	534	667	489	489	489	623	445	445	489	578	400	400	400
	7.94	934	667	667	623	890	623	623	578	801	578	578	578	756	534	534	489
	9.53	934	667	667	623	890	623	623	578	801	578	578	578	756	534	534	445
	12.70	1,824	1,290	1,290	1,112	1,735	1,157	1,157	1,023	1,601	1,068	1,068	1,023	1,512	979	979	845
	15.88	2,980	1,957	1,957	1,735	2,802	1,824	1,824	1,601	2,580	1,646	1,646	1,601	2,402	1,468	1,512	1,246
	19.05	4,493	2,891	2,891	2,491	4,226	2,580	2,669	2,269	3,914	2,180	2,402	2,269	3,648	1,868	2,180	1,646

Source: NDS 2015

Table A3.11 Selected bent plate Simpson™ connector reference strengths

Connector Type	Strength, lb (N) Load Direction (see Figure)	
	F_1	F_2
A34	340	340
	(1,512)	(1,512)
A35	510	510
	(2,269)	(2,269)
LTP4	500	500
	(2,224)	(2,224)
LTP5	470	470
	(2,091)	(2,091)
RBC	375	–
	(1,668)	–
GA1	200	200
	(890)	(890)
A21	365	175
	(1,624)	(778)

$$Z' = ZC_DC_MC_t$$

Notes: (1.) Values are for Spruce–Pine–Fir. Somewhat higher values are available for Douglas Fir or Southern Pine. (2.) Values are for floors. Increases are available for roof and wind/seismic conditions. (3.) Values assume all nail holes are filled

Source: Simpson Wood Construction Connectors Catalog 2015–2016

Table A3.12 Selected Simpson™ joist hanger reference strengths

		$Z' = ZC_DC_MC_t$		Allowable Load	
		Member Size	Connector Type	lb	(N)
Top Mount	Sawn Lumber	2 × 4	HU24TF	930	(4,137)
		2 × 6	JB26	815	(3,625)
		2 × 8	JB28	820	(3,648)
		2 × 10	JB210A	1,190	(5,293)
		2 × 12	JB212A	1,190	(5,293)
		2 × 14	JB214A	1,190	(5,293)
	Structural Composite Lumber	$1^3/_4 × 7^1/_4$	LBV1.81/7.25	2,060	(9,163)
		$1^3/_4 × 9^1/_4$	LBV1.81/9.25	2,060	(9,163)
		$1^3/_4 × 11^7/_8$	LBV1.81/11.88	2,060	(9,163)
		$1^3/_4 × 14$	LBV1.81/14	2,060	(9,163)
		$1^3/_4 × 16$	LBV1.81/16	2,060	(9,163)
	I-Joist	$2 × 9^1/_2$	ITS2.06/9.5	1,150	(5,115)
		$1^3/_4 × 11^7/_8$	IUS1.81/11.88	1,185	(5,271)
		$2 × 11^7/_8$	ITS2.06/11.88	1,150	(5,115)
		2 × 14	ITS2.06/14	1,150	(5,115)
		2 × 16	ITS2.06/16	1,150	(5,115)
Face Mount	Sawn Lumber	2 × 4	LU24	475	(2,113)
		2 × 6	LU26	740	(3,292)
		2 × 8	LU28	950	(4,226)
		2 × 10	LU210	1,190	(5,293)
		2 × 12	HU212	1,280	(5,694)
		2 × 14	HU214	1,540	(6,850)
	Structural Composite Lumber	$1^3/_4 × 5^1/_2$	HU1.81/5	2,050	(9,119)
		$1^3/_4 × 7^1/_4$	HU7	2,050	(9,119)
		$1^3/_4 × 9^1/_2$	HU9	3,075	(13,678)
		$1^3/_4 × 11^1/_4$	HU11	3,845	(17,103)
		$1^3/_4 × 14$	H14	4,615	(20,529)
	I-Joist	$2 × 9^1/_2$	IUS2.06/9.5	815	(3,625)
		$2 × 11^7/_8$	IUS2.06/11.88	1,020	(4,537)
		2 × 14	IUS2.06/14	1,425	(6,339)
		2 × 16	IUS2.06/16	1,630	(7,251)

Notes: (1.) Values are for Spruce–Pine–Fir. Higher values are available for Douglas Fir or Southern Pine. (2.) Values are for floors. Increases are available for snow and roof conditions. (3.) Values assume all nail holes are filled

Source: Simpson Wood Construction Connectors Catalog 2015–2016

Table A3.13 Selected Simpson™ hold down and strap reference strengths

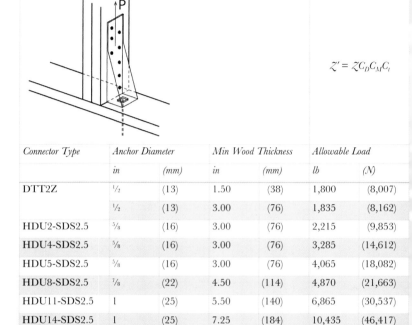

$$Z' = ZC_DC_MC_t$$

Connector Type	Anchor Diameter		Min Wood Thickness		Allowable Load	
	in	*(mm)*	*in*	*(mm)*	*lb*	*(N)*
DTT2Z	$\frac{1}{2}$	(13)	1.50	(38)	1,800	(8,007)
	$\frac{1}{2}$	(13)	3.00	(76)	1,835	(8,162)
HDU2-SDS2.5	$\frac{5}{8}$	(16)	3.00	(76)	2,215	(9,853)
HDU4-SDS2.5	$\frac{5}{8}$	(16)	3.00	(76)	3,285	(14,612)
HDU5-SDS2.5	$\frac{5}{8}$	(16)	3.00	(76)	4,065	(18,082)
HDU8-SDS2.5	$\frac{7}{8}$	(22)	4.50	(114)	4,870	(21,663)
HDU11-SDS2.5	1	(25)	5.50	(140)	6,865	(30,537)
HDU14-SDS2.5	1	(25)	7.25	(184)	10,435	(46,417)
LTT19	$\frac{1}{2}$	(13)	3.00	(76)	1,150	(5,115)
LTT20B	$\frac{1}{2}$	(13)	3.00	(76)	1,290	(5,738)
HTT4	$\frac{5}{8}$	(16)	3.00	(76)	3,640	(16,192)

Notes: (1.) Values are for Spruce–Pine–Fir. Higher values are available for Douglas Fir or Southern Pine. (2.) Values are for wind and seismic applications. (3.) Values assume all nail holes are filled

Source: Simpson Wood Construction Connectors Catalog 2015–2016

Table A3.14 Selected Simpson™ strap reference strengths

$$Z' = ZC_DC_MC_t$$

	Width		Length		Allowable Load	
	in	*(mm)*	*in*	*(mm)*	*lb*	*(N)*
LSTA9	1¼	*(32)*	9	*(229)*	*635*	*(2,825)*
LSTA12	1¼	(32)	12	(305)	795	(3,536)
LSTA15	1¼	(32)	15	(381)	950	(4,226)
LSTA18	1¼	(32)	18	(457)	1,110	(4,938)
LSTA21	1¼	(32)	21	(533)	1,235	(5,494)
LSTA24	1¼	(32)	24	(610)	1,235	(5,494)
MSTI26	2¹/₁₆	(52)	26	(660)	2,325	(10,342)
MSTI36	2¹/₁₆	(52)	36	(914)	3,220	(14,323)
MSTI48	2¹/₁₆	(52)	48	(1,219)	4,290	(19,083)
MSTI60	2¹/₁₆	(52)	60	(1,524)	5,080	(22,597)
MSTI72	2¹/₁₆	(52)	72	(1,829)	5,080	(22,597)
CS14	1¼	(32)	300	(7,620)	2,490	(11,076)
CS18	1¼	(32)	300	(7,620)	1,370	(6,094)
CS22	1¼	(32)	300	(7,620)	845	(3,759)

Notes: (1.) Values are for Spruce–Pine–Fir. Higher values are available for Douglas Fir or Southern Pine. (2.) Values are for wind and seismic applications. (3.) Values assume all nail holes are filled

Source: Simpson Wood Construction Connectors Catalog 2015–2016

Adjustment Factors

Appendix 4

Table A4.1 Load duration adjustment factors C_D

Load Duration	C_D	Load Type
Permanent	0.9	Dead
10 years	1.0	Occupancy live
2 months	1.15	Snow
7 days	1.25	Construction
10 minutes	1.60	Wind & earthquake
Impact	2.0	Impact

Source: NDS 2015

Table A4.2 Wood wet service adjustment factor C_M

Visually Graded Sawn Lumber					
F_b	F_t	F_v	$F_{c\perp}$	F_c	E, E_{min}
0.85	1.0	0.97	0.67	0.8	0.9

Visually Graded Timbers					
F_b	F_t	F_v	$F_{c\perp}$	F_c	E, E_{min}
1.00	1.00	1.00	0.67	0.91	1.00

Glue Laminated Timbers					
F_b	F_t	F_v	$F_{c\perp}$	F_c	E, E_{min}
0.8	0.8	0.875	0.53	0.73	0.833

Structural Composite Lumber
SCL is typically used in the dry condition ($C_M = 1.0$).
Consult the manufacturer if to be used wet

Source: NDS 2015

Table A4.3 Fastener wet service adjustment factor C_M

| Fastener Type | Moisture Content | | C_M |
	At Fabrication	In-Service	
Lateral Loads			
Dowel-type fasteners (bolts, lag screws, nails)	≤ 19%	≤ 19%	1.0
	> 19%	≤ 19%	0.4
	Any	> 19%	0.7
Withdrawal Loads			
Lag screws & wood screws	Any	< 19%	1.0
	Any	> 19%	0.7
Nails & spikes	≤ 19%	≤ 19%	1.0
	> 19%	≤ 19%	0.25
	≤ 19%	> 19%	0.25
	> 19%	> 19%	1.0

Notes: (1.) C_M = 0.7 for dowel-type fasteners with diameter D less than 1/4".
(2.) C_M = 1.0 for dowel-type fastener connections with: one fastener only, or two or more fasteners placed in a single row parallel to grain

Source: NDS 2015

Table A4.4 Temperature adjustment factors C_t

| Reference Values | Moisture Condition | C_t | | |
		$T ≤ 100°F$ $T ≤ 38°C$	$100°F < T ≤ 125°F$ $38°C < T ≤ 52°C$	$125°F < T ≤ 150°F$ $52°C < T ≤ 65°C$
Timber				
F_t, E, E_{min}	Wet or dry	1.0	0.9	0.9
F_b, F_v, F_c, F_c	Dry	1.0	0.8	0.7
	Wet	1.0	0.7	0.5
Connections				
	Dry	1.0	0.8	0.7
	Wet	1.0	0.7	0.5

Source: NDS 2015

Table A4.5 Size adjustment factor C_F

Grades	Depth		C_F		
	in	(mm)	F_b	F_t	F_c
Visually Graded, Sawn Lumber					
Select structural No. 1 & Btr No. 1, No. 2, No. 3	4	(102)	1.5	1.5	1.15
	5	(127)	1.4	1.4	1.1
	6	(153)	1.3	1.3	1.1
	8	(204)	1.2	1.2	1.05
	10	(254)	1.1	1.1	1.0
	12	(305)	1.0	1.0	1.0
	14	(356)	0.9	0.9	0.9
Stud	4	(102)	1.1	1.1	1.05
	6	(153)	1.0	1.0	1.0
	8	(204)	Use No. 3 grade design values and size factors		
Construction standard	4	(102)	1.0	1.0	1.0
Visually Graded Southern Pine Size factors are built into Table A2.2					
Machine Rated Lumber Size factors are built into Table A2.3					
Visually Graded Timbers					
	12	(305)	1.0	1.0	1.0
	16	(407)	0.969	1.0	1.0
	20	(508)	0.945	1.0	1.0
	24	(610)	0.926	1.0	1.0
	30	(762)	0.903	1.0	1.0
	36	(915)	0.885	1.0	1.0
	42	(1,067)	0.870	1.0	1.0
	48	(1,220)	0.857	1.0	1.0
	54	(1,372)	0.846	1.0	1.0
	60	(1,524)	0.836	1.0	1.0

Source: NDS 2015

Table A4.6 Glued laminated volume adjustment factor C_v

| Imperial | | Metric | | Length, ft (m) | | | | | | | | |
b (in)	d (in)	b (mm)	d (mm)	10 (3.05)	15 (4.58)	20 (6.10)	25 (7.62)	30 (9.15)	35 (10.67)	40 (12.20)	50 (15.24)	60 (18.29)
5½	12	(140)	(305)	1.00	1.00	1.00	0.976	0.958	0.944	0.931	0.910	0.894
	16		(406)	1.00	1.00	0.97	0.948	0.931	0.917	0.905	0.885	0.869
	20		(508)	1.00	0.976	0.948	0.927	0.910	0.897	0.885	0.865	0.849
	24		(610)	1.00	0.958	0.931	0.910	0.894	0.880	0.869	0.849	0.834
	28		(711)	0.983	0.944	0.917	0.897	0.880	0.867	0.855	0.836	0.821
	32		(813)	0.970	0.931	0.905	0.885	0.869	0.855	0.844	0.825	0.810
	36		(914)	0.958	0.920	0.894	0.874	0.858	0.845	0.834	0.816	0.801
6¾	18	(171)	(457)	1.00	0.966	0.939	0.918	0.901	0.888	0.876	0.857	0.841
	24		(610)	0.978	0.939	0.912	0.892	0.876	0.862	0.851	0.832	0.817
	30		(762)	0.956	0.918	0.892	0.872	0.857	0.843	0.832	0.814	0.799
	36		(914)	0.939	0.901	0.876	0.857	0.841	0.828	0.817	0.799	0.785
	42		(1,067)	0.924	0.888	0.862	0.843	0.828	0.816	0.805	0.787	0.773
	48		(1,219)	0.912	0.876	0.851	0.832	0.817	0.805	0.794	0.777	0.762
	54		(1,372)	0.901	0.866	0.841	0.823	0.808	0.795	0.785	0.767	0.754
	60		(1,524)	0.892	0.857	0.832	0.814	0.799	0.787	0.777	0.759	0.746

Imperial		Metric		Length, ft (m)								
b (in)	d (in)	b (mm)	d (mm)	10 (3.05)	15 (4.58)	20 (6.10)	25 (7.62)	30 (9.15)	35 (10.67)	40 (12.20)	50 (15.24)	60 (18.29)
8¼	24	(222)	(610)	0.953	0.915	0.889	0.869	0.853	0.840	0.829	0.811	0.796
	30		(762)	0.932	0.895	0.869	0.850	0.835	0.822	0.811	0.793	0.779
	36		(914)	0.915	0.878	0.853	0.835	0.820	0.807	0.796	0.779	0.765
	42		(1,067)	0.901	0.865	0.840	0.822	0.807	0.795	0.784	0.767	0.753
	48		(1,219)	0.889	0.853	0.829	0.811	0.796	0.784	0.774	0.757	0.743
	54		(1,372)	0.878	0.843	0.820	0.801	0.787	0.775	0.765	0.748	0.734
	60		(1,524)	0.869	0.835	0.811	0.793	0.779	0.767	0.757	0.740	0.727
10¾	30	(273)	(762)	0.913	0.876	0.851	0.833	0.818	0.805	0.794	0.777	0.763
	36		(914)	0.896	0.860	0.836	0.818	0.803	0.791	0.780	0.763	0.749
	42		(1,067)	0.882	0.847	0.823	0.805	0.791	0.778	0.768	0.751	0.738
	48		(1,219)	0.871	0.836	0.812	0.794	0.780	0.768	0.758	0.741	0.728
	54		(1,372)	0.860	0.826	0.803	0.785	0.771	0.759	0.749	0.733	0.719
	60		(1,524)	0.851	0.818	0.794	0.777	0.763	0.751	0.741	0.725	0.712

Note: Values are conservative for Southern Pine

Source: NDS 2015

Table A4.7 Structural composite Lumber (SCL) volumne adjustment factor C_V

Material	Member Depth, in (mm)													
	3.5 (89)	5.5 (140)	7.25 (184)	9.25 (235)	9.50 (241)	11.88 (302)	14.0 (356)	16.0 (406)	18.0 (457)	20.0 (508)	24.0 (610)	48.0 (1,219)	54.0 (1372)
	C_v Adjustment Factor												
LVL	1.18	1.11	1.07	1.04	1.03	1.00	0.98	0.96	0.95	0.93	0.91	0.83	—
PSL	1.15	1.09	1.06	1.03	1.03	1.00	0.98	0.97	0.96	0.94	0.93	0.86	0.85
LSL	1.12	1.07	1.05	1.02	1.02	1.00	0.99	0.97	0.96	0.95	0.94	0.88	—

Source: ICC ESR-1387

Table A4.8 Flat use adjustment factor C_{fu}

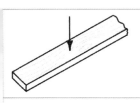

Sawn Lumber

Depth		Thickness	
in	*(mm)*	*2 & 3 in (51 & 76 mm)*	*4 in (102 mm)*
3	(77)	1.0	–
4	(102)	1.1	1.0
5	(127)	1.1	1.05
6	(153)	1.15	1.05
8	(204)	1.15	1.05
10+	(254+)	1.2	1.1

Structural Composite Lumber
See reference values listed in actual ICC-ESR report

Visually Graded Timbers 5 × 5

Grade	F_b	E and E_{min}	Other Properties
Select Structural	0.86	1.00	1.00
No. 1	0.74	0.90	1.00
No. 2	1.00	1.00	1.00

Gluelaminated Timbers

Width		
in	*(mm)*	
2½	(64)	1.19
3⅛	(80)	1.16
5½	(140)	1.10
6¾	(172)	1.07
8¾	(223)	1.04
10¾	(274)	1.01

Source: NDS 2015

Table A4.9 Curvature adjustment factor C_c

Radius of Curvature		Lamination Thickness, in (mm)		
in	(mm)	0.75 (19)	1 (25)	1.5 (38)
80	2,032	0.872		
120	3,048	0.922	0.872	
150	3,810	0.950	0.911	0.872
200	5,080	0.972	0.950	0.888
250	6,350	0.982	0.968	0.928
300	7,620	0.988	0.978	0.950
400	10,160	0.993	0.988	0.972
500	12,700	0.996	0.992	0.982
750	19,050	0.998	0.996	0.992
1,000	25,400	0.999	0.998	0.996
1,500	38,100	1.000	0.999	0.998
2,000	50,800	1.000	1.000	0.999
>3,000	>76,200	1.000	1.000	1.000

Source: NDS 2015

Appendix 4

Table A4.10 Incising adjustment factor C_i

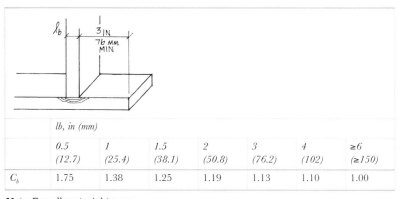

Design Value	C_i
E, E_{min}	0.95
F_b, F_t, F_c, F_v	0.80
$F_{c\perp}$	1.00

Source: NDS 2015

Table A4.11 Bearing area adjustment factor C_b

	lb, in (mm)						
	0.5 (12.7)	1 (25.4)	1.5 (38.1)	2 (50.8)	3 (76.2)	4 (102)	≥6 (≥150)
C_b	1.75	1.38	1.25	1.19	1.13	1.10	1.00

Note: For all material types
Source: NDS 2015

Table A4.12 Group action factors C_g for bolt or lag screws

A_s/A_m	A_s		Fasteners in a Row						
	in^2	(mm^2)	2	3	4	5	6	7	8
Wood Main & Side Members									
0.5	5	(127)	0.98	0.92	0.84	0.75	0.68	0.61	0.55
	12	(305)	0.99	0.96	0.92	0.87	0.81	0.76	0.70
	20	(508)	0.99	0.98	0.95	0.91	0.87	0.83	0.78
	40	(1,016)	1.00	0.99	0.97	0.95	0.93	0.90	0.87
1	5	(127)	1.00	0.97	0.91	0.85	0.78	0.71	0.64
	12	(305)	1.00	0.99	0.96	0.93	0.88	0.84	0.79
	20	(508)	1.00	0.99	0.98	0.95	0.92	0.89	0.86
	40	(1,016)	1.00	1.00	0.99	0.98	0.96	0.94	0.92
Steel Side & Wood Main Members									
12	5	(127)	0.97	0.89	0.80	0.70	0.62	0.55	0.49
	8	(204)	0.98	0.93	0.85	0.77	0.70	0.63	0.57
	16	(407)	0.99	0.96	0.92	0.86	0.80	0.75	0.69
	24	(610)	0.99	0.97	0.94	0.90	0.85	0.81	0.76
	40	(1,016)	1.00	0.98	0.96	0.94	0.90	0.87	0.83
	64	(1,626)	1.00	0.99	0.98	0.96	0.94	0.91	0.88
	120	(3,048)	1.00	0.99	0.99	0.98	0.96	0.95	0.93
	200	(5,080)	1.00	1.00	0.99	0.99	0.98	0.97	0.96

Notes: Factors are conservative for $D < 1$ in (25 mm), $s < 4$ in (100 mm), $E_{wood} > 1,400,000$ lb/in^2 (9,652,700 kN/m^2), $E_{steel} > 29,000,000$ lb/in^2 (199,948,000 kN/m^2), or $A_m/A_s > 12$

Source: NDS 2015

Table A4.13 End distance-based geometry adjustment factor C_Δ

END DISTANCE ‖ TO END GRAIN

END DISTANCE ⊥ TO GRAIN

Diameter D		C_Δ		C_Δ		C_Δ		C_Δ		C_Δ	
(in)	(mm)	2.0D		2.5D		3.0D		3.5D		4.0D	
End Distance, in (mm)											
2D to 4D—End distances for varying diameters											
0.250	(6.4)	0.50	(12.7)	0.63	(15.9)	0.75	(19.1)	0.88	(22.2)	1.00	(25.4)
0.375	(9.5)	0.75	(19.1)	0.94	(23.8)	1.13	(28.6)	1.31	(33.3)	1.50	(38.1)
0.500	(12.7)	1.00	(25.4)	1.25	(31.8)	1.50	(38.1)	1.75	(44.5)	2.00	(50.8)
0.625	(15.9)	1.25	(31.8)	1.56	(39.7)	1.88	(47.6)	2.19	(55.6)	2.50	(63.5)
0.750	(19.1)	1.50	(38.1)	1.88	(47.6)	2.25	(57.2)	2.63	(66.7)	3.00	(76.2)
3.5D to 7D—End distances for varying diameters											
0.250	(6.4)	0.88	(22.2)	1.00	(25.4)	1.25	(31.8)	1.50	(38.1)	1.75	(44.5)
0.375	(9.5)	1.31	(33.3)	1.50	(38.1)	1.88	(47.6)	2.25	(57.2)	2.63	(66.7)
0.500	(12.7)	1.75	(44.5)	2.00	(50.8)	2.50	(63.5)	3.00	(76.2)	3.50	(88.9)
0.625	(15.9)	2.19	(55.6)	2.50	(63.5)	3.13	(79.4)	3.75	(95.3)	4.38	(111.1)
0.750	(19.1)	2.63	(66.7)	3.00	(76.2)	3.75	(95.3)	4.50	(114.3)	5.25	(133.4)

Table A4.14 Spacing based geometry adjustment factor C_Δ

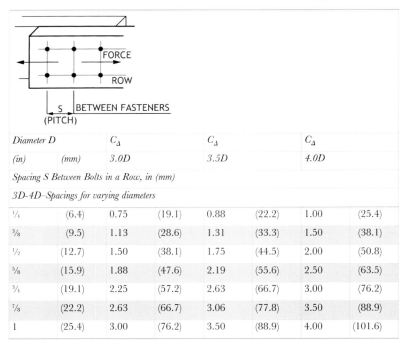

Diameter D		C_Δ		C_Δ		C_Δ	
(in)	*(mm)*	*3.0D*		*3.5D*		*4.0D*	
Spacing S Between Bolts in a Row, in (mm)							
3D-4D–Spacings for varying diameters							
1/4	(6.4)	0.75	(19.1)	0.88	(22.2)	1.00	(25.4)
3/8	(9.5)	1.13	(28.6)	1.31	(33.3)	1.50	(38.1)
1/2	(12.7)	1.50	(38.1)	1.75	(44.5)	2.00	(50.8)
5/8	(15.9)	1.88	(47.6)	2.19	(55.6)	2.50	(63.5)
3/4	(19.1)	2.25	(57.2)	2.63	(66.7)	3.00	(76.2)
7/8	(22.2)	2.63	(66.7)	3.06	(77.8)	3.50	(88.9)
1	(25.4)	3.00	(76.2)	3.50	(88.9)	4.00	(101.6)

Table A4.15 Endgrain adjustment factor C_{eg}

Action	C_{eg}		
	Lag Screw	*Wood Screw*	*Nail & Spike*
Lateral	0.67	0.67	0.67
Withdrawal	0.75	0	0

Source: NDS 2015

Table A4.16 Toe-nail adjustment factor C_{tn}

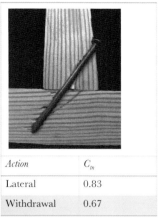

Action	C_{tn}
Lateral	0.83
Withdrawal	0.67

Source: NDS 2015

Table A4.17 Format conversion factor K_F

Application	Property	K_F
Member	F_b	2.54
	F_t	2.70
	F_v, F_{rt}, F_s	2.88
	F_c	2.40
	$F_{c\perp}$	1.67
	E_{min}	1.76
Connections	All	3.32

Note: For LRFD only
Source: NDS 2015

Table A4.18 Resistance factor ϕ

Application	Property	Symbol	Value
Member	F_b	ϕ_b	0.85
	F_t	ϕ_t	0.80
	F_v, F_{rt}, F_s	ϕ_v	0.75
	F_c, $F_{c\perp}$	ϕ_c	0.90
	E_{min}	ϕ_s	0.85
Connections	All	ϕ_z	0.65

Note: For LRFD only
Source: NDS 2015

Table A4.19 Time effect factor λ

Load Combination	λ
1.4D	0.6
1.2D + 1.6L + 0.5S	0.7 when L is from storage 0.8 when L is from occupancy 1.25 when L is from impact
1.2D + 1.6S + L	0.8
1.2D + 1.0W + L + 0.5S	1.0
1.2D + 1.0E + L + 0.2S	1.0
0.9D + 1.0W	1.0
0.9D + 1.0E	1.0

Note: For LRFD only
Source: NDS 2015

Simple Design Aids

Appendix 5

Table A5.1 Joist span table, normal duration loading, L/360 LL Defl limit

10psf Dead + 30psf Live						Joist size 2 × 8			Joist size 2 × 10			Joist size 2 × 12		
Species & Grade		Spacing		Span				Span				Span		
		in	(mm)	ft	in	(m)	ft	in	(m)	ft	in	(m)		
Hem-Fir	SS	12	(300)	15.5	15 6	(4.7)	19.84	19 10	(6.0)	24.12	24 1	(7.4)		
		16	(400)	14.1	14 1	(4.3)	18.02	18 0	(5.5)	21.92	21 11	(6.7)		
		19.2	(480)	13.3	13 3	(4.1)	16.96	16 11	(5.2)	20.63	20 7	(6.3)		
	#1	12	(300)	15.2	15 2	(4.6)	19.41	19 4	(5.9)	23.61	23 7	(7.2)		
		16	(400)	13.6	13 6	(4.1)	17.31	17 3	(5.3)	21.06	21 0	(6.4)		
		19.2	(480)	12.4	12 4	(3.8)	15.81	15 9	(4.8)	19.22	19 2	(5.9)		
	#2	12	(300)	14.6	14 7	(4.5)	18.67	18 8	(5.7)	22.70	22 8	(6.9)		
		16	(400)	12.7	12 8	(3.9)	16.17	16 2	(4.9)	19.66	19 7	(6.0)		
		19.2	(480)	11.6	11 6	(3.5)	14.76	14 9	(4.5)	17.95	17 11	(5.5)		
S-P-F	SS	12	(300)	15.2	15 2	(4.6)	19.41	19 4	(5.9)	23.61	23 7	(7.2)		
		16	(400)	13.8	13 9	(4.2)	17.64	17 7	(5.4)	21.45	21 5	(6.5)		
		19.2	(480)	13.0	13 0	(4.0)	16.60	16 7	(5.1)	20.19	20 2	(6.2)		
	#1/#2	12	(300)	14.8	14 10	(4.5)	18.94	18 11	(5.8)	23.04	23 0	(7.0)		
		16	(400)	12.9	12 10	(3.9)	16.40	16 4	(5.0)	19.95	19 11	(6.1)		
		19.2	(480)	11.7	11 8	(3.6)	14.97	14 11	(4.6)	18.21	18 2	(5.6)		
	#3	12	(300)	11.2	11 2	(3.4)	14.32	14 3	(4.4)	17.41	17 4	(5.3)		
		16	(400)	9.7	9 8	(3.0)	12.40	12 4	(3.8)	15.08	15 0	(4.6)		
		19.2	(480)	8.9	8 10	(2.7)	11.32	11 3	(3.4)	13.77	13 9	(4.2)		
So. Pine	SS	12	(300)	16.2	16 2	(4.9)	20.63	20 7	(6.3)	25.09	25 1	(7.6)		
		16	(400)	14.7	14 8	(4.5)	18.74	18 8	(5.7)	22.80	22 9	(6.9)		
		19.2	(480)	13.8	13 9	(4.2)	17.64	17 7	(5.4)	21.45	21 5	(6.5)		
	#1	12	(300)	15.5	15 6	(4.7)	19.84	19 10	(6.0)	24.12	24 1	(7.4)		
		16	(400)	14.1	14 1	(4.3)	18.02	18 0	(5.5)	21.92	21 11	(6.7)		
		19.2	(480)	13.3	13 3	(4.1)	16.96	16 11	(5.2)	20.63	20 7	(6.3)		
	#2	12	(300)	14.9	14 10	(4.5)	18.97	18 11	(5.8)	23.07	23 0	(7.0)		
		16	(400)	13.2	13 2	(4.0)	16.87	16 10	(5.1)	20.51	20 6	(6.3)		
		19.2	(480)	12.1	12 0	(3.7)	15.40	15 4	(4.7)	18.72	18 8	(5.7)		

10psf Dead + 40psf Live		Spacing		Joist size 2 × 8			Joist size 2 × 10			Joist size 2 × 12		
Species & Grade				Span			Span			Span		
		in	(mm)	ft	in	(m)	ft	in	(m)	ft	in	(m)
Hem-Fir	SS	12	(300)	14.1	14 1	(4.3)	18.02	18 0	(5.5)	21.92	21 11	(6.7)
		16	(400)	12.8	12 10	(3.9)	16.37	16 4	(5.0)	19.91	19 10	(6.1)
		19.2	(480)	12.1	12 0	(3.7)	15.41	15 4	(4.7)	18.74	18 8	(5.7)
	#1	12	(300)	13.8	13 9	(4.2)	17.64	17 7	(5.4)	21.45	21 5	(6.5)
		16	(400)	12.1	12 1	(3.7)	15.49	15 5	(4.7)	18.84	18 10	(5.7)
		19.2	(480)	11.1	11 0	(3.4)	14.14	14 1	(4.3)	17.19	17 2	(5.2)
	#2	12	(300)	13.1	13 1	(4.0)	16.70	16 8	(5.1)	20.31	20 3	(6.2)
		16	(400)	11.3	11 4	(3.5)	14.46	14 5	(4.4)	17.59	17 7	(5.4)
		19.2	(480)	10.3	10 4	(3.2)	13.20	13 2	(4.0)	16.05	16 0	(4.9)
S-P-F	SS	12	(300)	13.8	13 9	(4.2)	17.64	17 7	(5.4)	21.45	21 5	(6.5)
		16	(400)	12.6	12 6	(3.8)	16.03	16 0	(4.9)	19.49	19 5	(5.9)
		19.2	(480)	11.8	11 9	(3.6)	15.08	15 0	(4.6)	18.34	18 4	(5.6)
	#1/ #2	12	(300)	13.3	13 3	(4.0)	16.94	16 11	(5.2)	20.60	20 7	(6.3)
		16	(400)	11.5	11 5	(3.5)	14.67	14 8	(4.5)	17.84	17 10	(5.4)
		19.2	(480)	10.5	10 5	(3.2)	13.39	13 4	(4.1)	16.29	16 3	(5.0)
	#3	12	(300)	10.0	10 0	(3.1)	12.81	12 9	(3.9)	15.57	15 6	(4.7)
		16	(400)	8.7	8 8	(2.6)	11.09	11 1	(3.4)	13.49	13 5	(4.1)
		19.2	(480)	7.9	7 11	(2.4)	10.12	10 1	(3.1)	12.31	12 3	(3.8)
So. Pine	SS	12	(300)	14.7	14 8	(4.5)	18.74	18 8	(5.7)	22.80	22 9	(6.9)
		16	(400)	13.3	13 4	(4.1)	17.03	17 0	(5.2)	20.71	20 8	(6.3)
		19.2	(480)	12.6	12 6	(3.8)	16.03	16 0	(4.9)	19.49	19 5	(5.9)
	#1	12	(300)	14.1	14 1	(4.3)	18.02	18 0	(5.5)	21.92	21 11	(6.7)
		16	(400)	12.8	12 10	(3.9)	16.37	16 4	(5.0)	19.91	19 10	(6.1)
		19.2	(480)	12.1	12 0	(3.7)	15.41	15 4	(4.7)	18.74	18 8	(5.7)
	#2	12	(300)	13.5	13 6	(4.1)	17.24	17 2	(5.3)	20.96	20 11	(6.4)
		16	(400)	11.8	11 9	(3.6)	15.08	15 1	(4.6)	18.35	18 4	(5.6)
		19.2	(480)	10.8	10 9	(3.3)	13.77	13 9	(4.2)	16.75	16 8	(5.1)

10psf Dead + 50psf Live				Joist size 2 × 8			Joist size 2 × 10			Joist size 2 × 12		
Species & Grade		Spacing		Span			Span			Span		
		in	(mm)	ft	in	(m)	ft	in	(m)	ft	in	(m)
Hem-Fir	SS	12	(300)	13.1	13 1	(4.0)	16.73	16 8	(5.1)	20.35	20 4	(6.2)
		16	(400)	11.9	11 10	(3.6)	15.20	15 2	(4.6)	18.49	18 5	(5.6)
		19.2	(480)	11.2	11 2	(3.4)	14.30	14 3	(4.4)	17.40	17 4	(5.3)
	#1	12	(300)	12.8	12 9	(3.9)	16.32	16 3	(5.0)	19.85	19 10	(6.1)
		16	(400)	11.1	11 0	(3.4)	14.14	14 1	(4.3)	17.19	17 2	(5.2)
		19.2	(480)	10.1	10 1	(3.1)	12.91	12 10	(3.9)	15.70	15 8	(4.8)
	#2	12	(300)	11.9	11 11	(3.6)	15.24	15 2	(4.6)	18.54	18 6	(5.7)
		16	(400)	10.3	10 4	(3.2)	13.20	13 2	(4.0)	16.05	16 0	(4.9)
		19.2	(480)	9.4	9 5	(2.9)	12.05	12 0	(3.7)	14.66	14 7	(4.5)
S-P-F	SS	12	(300)	12.8	12 10	(3.9)	16.37	16 4	(5.0)	19.91	19 10	(6.1)
		16	(400)	11.7	11 7	(3.6)	14.88	14 10	(4.5)	18.09	18 1	(5.5)
		19.2	(480)	11.0	10 11	(3.3)	14.00	13 11	(4.3)	17.03	17 0	(5.2)
	#1/ #2	12	(300)	12.1	12 1	(3.7)	15.46	15 5	(4.7)	18.81	18 9	(5.7)
		16	(400)	10.5	10 5	(3.2)	13.39	13 4	(4.1)	16.29	16 3	(5.0)
		19.2	(480)	9.6	9 6	(2.9)	12.23	12 2	(3.7)	14.87	14 10	(4.5)
	#3	12	(300)	9.2	9 1	(2.8)	11.69	11 8	(3.6)	14.22	14 2	(4.3)
		16	(400)	7.9	7 11	(2.4)	10.12	10 1	(3.1)	12.31	12 3	(3.8)
		19.2	(480)	7.2	7 2	(2.2)	9.24	9 2	(2.8)	11.24	11 2	(3.4)
So. Pine	SS	12	(300)	13.6	13 7	(4.2)	17.40	17 4	(5.3)	21.16	21 1	(6.4)
		16	(400)	12.4	12 4	(3.8)	15.81	15 9	(4.8)	19.23	19 2	(5.9)
		19.2	(480)	11.7	11 7	(3.6)	14.88	14 10	(4.5)	18.09	18 1	(5.5)
	#1	12	(300)	13.1	13 1	(4.0)	16.73	16 8	(5.1)	20.35	20 4	(6.2)
		16	(400)	11.9	11 10	(3.6)	15.20	15 2	(4.6)	18.49	18 5	(5.6)
		19.2	(480)	11.2	11 2	(3.4)	14.30	14 3	(4.4)	17.40	17 4	(5.3)
	#2	12	(300)	12.5	12 5	(3.8)	15.90	15 10	(4.8)	19.34	19 4	(5.9)
		16	(400)	10.8	10 9	(3.3)	13.77	13 9	(4.2)	16.75	16 8	(5.1)
		19.2	(480)	9.9	9 10	(3.0)	12.57	12 6	(3.8)	15.29	15 3	(4.7)

15psf Dead + 100psf Live					Joist size 2 × 8			Joist size 2 × 10			Joist size 2 × 12		
Species & Grade		Spacing		Span			Span			Span			
		in	(mm)	ft	in	(m)	ft	in	(m)	ft	in	(m)	
Hem -Fir	SS	12	(300)	10.4	10 4	(3.2)	13.28	13 3	(4.0)	16.15	16 1	(4.9)	
		16	(400)	9.5	9 5	(2.9)	12.06	12 0	(3.7)	14.67	14 8	(4.5)	
		19.2	(480)	8.8	8 9	(2.7)	11.17	11 2	(3.4)	13.59	13 7	(4.1)	
	#1	12	(300)	9.2	9 2	(2.8)	11.79	11 9	(3.6)	14.34	14 4	(4.4)	
		16	(400)	8.0	8 0	(2.4)	10.21	10 2	(3.1)	12.42	12 5	(3.8)	
		19.2	(480)	7.3	7 3	(2.2)	9.32	9 3	(2.8)	11.34	11 4	(3.5)	
	#2	12	(300)	8.6	8 7	(2.6)	11.01	11 0	(3.4)	13.39	13 4	(4.1)	
		16	(400)	7.5	7 5	(2.3)	9.53	9 6	(2.9)	11.60	11 7	(3.5)	
		19.2	(480)	6.8	6 9	(2.1)	8.70	8 8	(2.7)	10.59	10 7	(3.2)	
S-P-F	SS	12	(300)	10.2	10 2	(3.1)	13.00	12 11	(4.0)	15.81	15 9	(4.8)	
		16	(400)	9.1	9 0	(2.8)	11.56	11 6	(3.5)	14.06	14 0	(4.3)	
		19.2	(480)	8.3	8 3	(2.5)	10.56	10 6	(3.2)	12.84	12 10	(3.9)	
	#1/ #2	12	(300)	8.8	8 9	(2.7)	11.17	11 2	(3.4)	13.59	13 7	(4.1)	
		16	(400)	7.6	7 6	(2.3)	9.67	9 8	(2.9)	11.77	11 9	(3.6)	
		19.2	(480)	6.9	6 11	(2.1)	8.83	8 9	(2.7)	10.74	10 8	(3.3)	
	#3	12	(300)	6.6	6 7	(2.0)	8.44	8 5	(2.6)	10.27	10 3	(3.1)	
		16	(400)	5.7	5 8	(1.7)	7.31	7 3	(2.2)	8.89	8 10	(2.7)	
		19.2	(480)	5.2	5 2	(1.6)	6.68	6 8	(2.0)	8.12	8 1	(2.5)	
So. Pine	SS	12	(300)	10.8	10 9	(3.3)	13.81	13 9	(4.2)	16.80	16 9	(5.1)	
		16	(400)	9.8	9 10	(3.0)	12.55	12 6	(3.8)	15.26	15 3	(4.7)	
		19.2	(480)	9.3	9 3	(2.8)	11.81	11 9	(3.6)	14.36	14 4	(4.4)	
	#1	12	(300)	10.4	10 4	(3.2)	13.28	13 3	(4.0)	16.15	16 1	(4.9)	
		16	(400)	9.1	9 0	(2.8)	11.56	11 6	(3.5)	14.06	14 0	(4.3)	
		19.2	(480)	8.3	8 3	(2.5)	10.56	10 6	(3.2)	12.84	12 10	(3.9)	
	#2	12	(300)	9.0	9 0	(2.7)	11.49	11 5	(3.5)	13.97	13 11	(4.3)	
		16	(400)	7.8	7 9	(2.4)	9.95	9 11	(3.0)	12.10	12 1	(3.7)	
		19.2	(480)	7.1	7 1	(2.2)	9.08	9 0	(2.8)	11.04	11 0	(3.4)	

Beam Solutions

Appendix 6

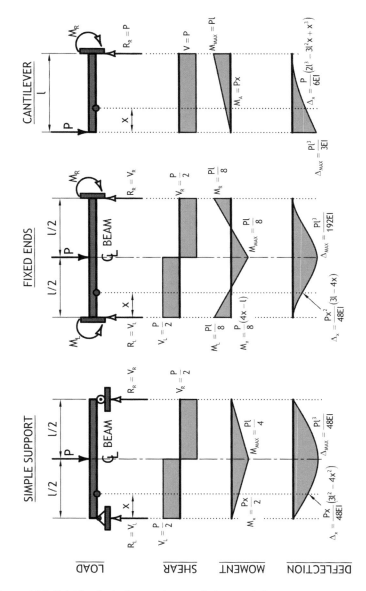

Figure A6.1 Point load, single-span beam solutions and diagrams

Appendix 6

331

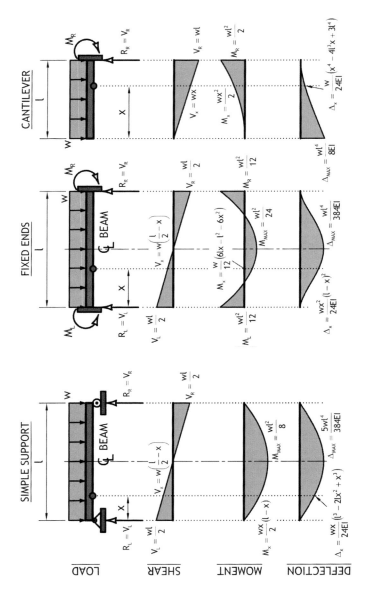

Figure A6.2 Uniform distributed load, single-span beam solutions and diagrams

Appendix 6

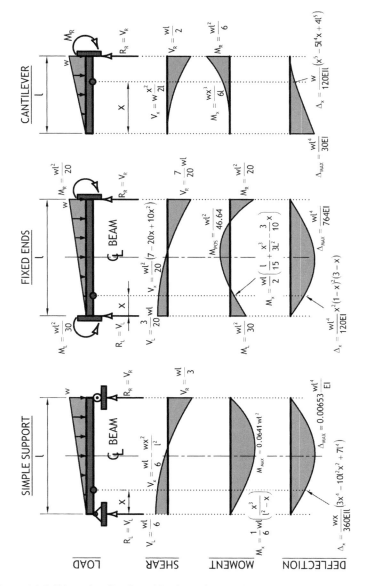

Figure A6.3 Triangular distributed load, single-span beam solutions and diagrams

Appendix 6

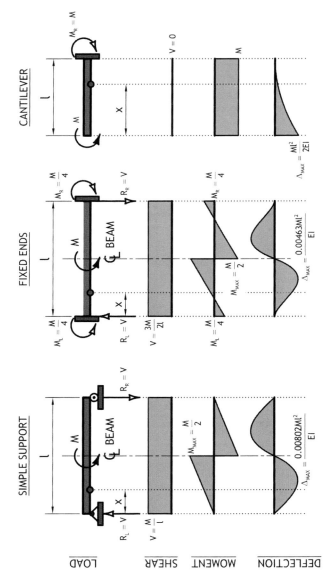

Figure A6.4 Moment load, single-span beam solutions and diagrams

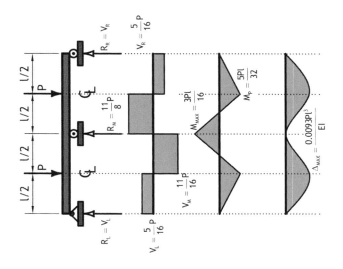

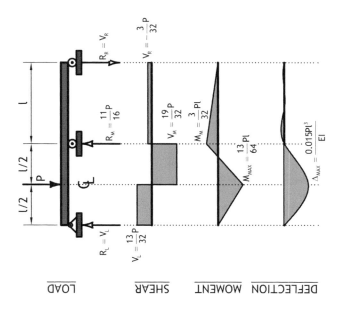

Figure A6.5 Point load, double-span beam solutions and diagrams

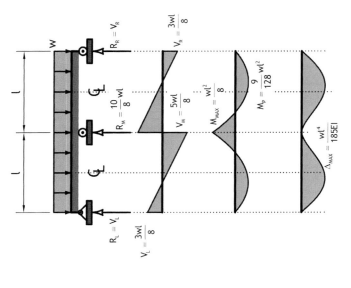

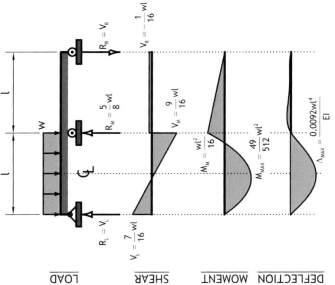

Figure A6.6 Uniform distributed load, double-span beam solutions and diagrams

List of Units

Appendix 7

Table A7.1 Units

Imperial Units	*Definition*	*Typical Use*
°	degrees	angle
deg	degrees	angle
ft	feet	length
ft^2	square feet	area
ft^3	cubic feet	volume
hr	hour	time
in	inches	length
in^2	square inch	area
in^3	cubic inch	volume
in^4	inches to the fourth power	moment of inertia
k	kip (1000 pounds)	force
k/ft	kips per foot (aka klf)	distributed linear force
k/ft^2	kips per square foot (aka ksf)	distributed area force, pressure
k/ft^3	kips per cubic foot (aka kcf)	density
k/in^2	kips per square inch (aka ksi)	distributed area force, pressure
k-ft	kip-feet	moment, torque
lb, lb_f	pound	force
lb/ft	pounds per foot (aka plf)	distributed linear force
lb/ft^2	pounds per square foot (aka psf)	distributed area force, pressure
lb/ft^3	pounds per cubic foot (aka pcf)	density
lb/in^2	pounds per square inch (aka psi)	distributed area force, pressure
lb-ft	pound-feet	moment, torque
rad	radian	angle
yd^3	cubic yard	volume

Metric Units	Definition	Typical Use
°	degrees	angle
deg	degrees	angle
g	gram	mass
hr	hour	time
kM/h	kilometers per hour	speed
kN	newton	force
kN	kiloNewton	force
kN/m	kiloNewton per meter	distributed linear force
kN/m^2	kiloNewton per square meter (aka kPa)	distributed area force, pressure
kN/m^3	kiloNewton per cubic foot	density
kN-m	kiloNewton-meter	moment, torque
m	meters	length
m^2	square meters	area
m^3	cubic meters	volume
min	minute	time
mm	millimeters	length
mm^2	square millimeters	area
mm^3	cubic millimeters	volume
mm^4	millimeters to the fourth power	moment of inertia
MN/m^2	kiloNewton per square inch (aka GPa)	distributed area force, pressure
N	newton	force
N/m	newtons per meter	distributed linear force
N/m^2	newtons per square meter (aka Pa)	distributed area force, pressure
N/m^3	newtons per cubic meter	density
N/mm^2	newtons per square millimeter (aka MPa)	distributed area force, pressure
Pa	newton per square meter (N/m^2)	distributed area force, pressure
rad	radian	angle

List of Symbols

Appendix 8

Table A8.1 Symbols

Symbol	Definition	Imperial	Metric
$\#_b$	bending (flexure) property or action	vary	
$\#_c$	compression property or action	vary	
$\#_D$	dead load-related action	vary	
$\#_H$	action in horizontal direction	lb or k	N, kN, MN
$\#_L$	live load related action	vary	
$\#_L$	action on left	vary	
$\#_n$	nominal capacity	vary	
$\#_R$	action on right	vary	
$\#_S$	snow load-related action	vary	
$\#_t$	tension property or action	vary	
$\#_u$	factored load, any type	vary	
$\#_v$	shear property or action	vary	
$\#_V$	action in vertical direction	lb or k	N, kN, MN
$\#_W$	wind load-related action	vary	
A	gross cross-section area	in^2	mm^2
A_{brg}	bearing area	in^2	mm^2
A_{eff}	effective cross-section area	in^2	mm^2
A_m	gross cross-section area of main member	in^2	mm^2
A_{net}	net cross-section area	in^2	mm^2
A_s	gross cross-section area of side member	in^2	mm^2
b	width (thickness) of bending member	in	mm
b_s	shear wall segment width	ft	m
c	distance to neutral axis from tension or compression face	in	mm
c	column buckling factor	unitless	
C	compression, concrete compression resultant	lb or k	N, kN, MN
C_b	bearing area factor	unitless	
C_c	curvature factor	unitless	
C_D	load duration factor	unitless	

Symbol	Definition	Imperial	Metric
C_Δ	geometry factor	unitless	
C_{di}	diaphragm factor	unitless	
C_{eg}	end grain factor	unitless	
C_F	size factor	unitless	
C_{fu}	flat use factor	unitless	
C_g	group action factor	unitless	
C_I	stress interaction factor	unitless	
C_i	incising factor	unitless	
C_L	beam stability factor	unitless	
C_M	wet service factor	unitless	
C_P	column stability factor	unitless	
C_r	repetitive member factor	unitless	
C_t	temperature factor	unitless	
C_T	buckling stiffness factor	unitless	
C_{tn}	toe-nail factor	unitless	
C_V	volume factor	unitless	
C_{vr}	shear reduction factor	unitless	
d	depth (height) of bending member	in	mm
d	smallest dimension of compression member	in	mm
d	pennyweight of nail or spike	unitless	
D	dowel-type fastener diameter	in	mm
D	dead load	k, k/ft, k/ft^2, lb, lb/ft, lb/ft^2	kN, kN/m, kN/m^2, N, N/m, N/m^2
δ	deflection	in	mm
δ_{LT}	deflection due to long-term loads	in	mm
δ_{ST}	deflection due to short-term loads	in	mm
δ_T	total deflection from short- and long-term loads	in	mm
E	reference modulus of elasticity	lb/in^2	GN/m^2, GPa
E	seismic load	lb or k	N, kN, MN

Symbol	Definition	Imperial	Metric
ϵ	strain	unitless	
e	eccentricity	in	mm
E'	adjusted modulus of elasticity	lb/in²	GN/m², GPa
E_{min}	reference modulus of elasticity for beam and column stability	lb/in²	GN/m², GPa
E'_{min}	adjusted modulus of elasticity for beam and column stability	lb/in²	GN/m², GPa
ϕ	resistance factor		
f	stress	lb/in²	MN/m², MPa
F^*_b	reference bending strength multiplied by all adjustment factors but C_L	lb/in²	MN/m², MPa
F^*_c	reference compression strength multiplied by all adjustment factors but C_P	lb/in²	MN/m², MPa
f_b	bending stress	lb/in²	MN/m², MPa
F_b	reference bending design stress	lb/in²	MN/m², MPa
F'_b	adjusted bending design stress	lb/in²	MN/m², MPa
F_{bE}	critical buckling stress for bending	lb/in²	MN/m², MPa
F_c	compression stress parallel to grain	lb/in²	MN/m², MPa
F_c	reference compression design stress parallel to grain	lb/in²	MN/m², MPa
F'_c	adjusted compression design stress parallel to grain	lb/in²	MN/m², MPa
$F_{c\perp}$	reference compression design stress perpendicular to grain	lb/in²	MN/m², MPa
$f_{c\perp}$	compression stress perpendicular to grain	lb/in²	MN/m², MPa
$F'_{c\perp}$	adjusted compression design stress perpendicular to grain	lb/in²	MN/m², MPa
F_{cE}	critical buckling strength for compression	lb/in²	MN/m², MPa
F_e	dowel bearing strength	lb/in²	MN/m², MPa
$F_{e\perp}$	dowel bearing strength perpendicular to grain	lb/in²	MN/m², MPa
$F_{e\parallel}$	dowel bearing strength parallel to grain	lb/in²	MN/m², MPa

Symbol	Definition	Imperial	Metric
f_t	tension stress parallel to grain	lb/in^2	MN/m^2, MPa
F_t	reference tension design stress	lb/in^2	MN/m^2, MPa
F'_t	adjusted tension design stress	lb/in^2	MN/m^2, MPa
f_v	shear stress parallel to grain	lb/in^2	MN/m^2, MPa
F_v	reference shear design stress	lb/in^2	MN/m^2, MPa
F'_v	adjusted shear design stress	lb/in^2	MN/m^2, MPa
γ	unit weight (density)	lb/ft^3	kN/m^3
H	soil load	k, k/ft, k/ft^2, lb, lb/ft, lb/ft^2	kN, kN/m, kN/m^2, N, N/m, N/m^2
h	section height or depth, wall height	ft, in	m, mm
h_w	wall height	ft, in	m, mm
h_x	maximum spacing between lengthwise bars in column without cross-tie	ft, in	m, mm
I, I_x, I_y	moment of inertia	in^4	mm^4
k	effective length factor	unitless	
K_{cr}	creep factor	unitless	
K_F	format conversion factor		
λ	time effect factor		
L	live load	k, k/ft, k/ft^2, lb, lb/ft, lb/ft^2	kN, kN/m, kN/m^2, N, N/m, N/m^2
l	span length	in or ft	m, mm
L, L'	length or height	ft	m
l_b	bearing length	in or ft	mm
l_c	clear span	in or ft	m, mm
l_e	effective length	in	mm
L_o	base live load	k, k/ft, k/ft^2, lb, lb/ft, lb/ft^2	kN, kN/m, kN/m^2, N, N/m, N/m^2
l_t	tributary width	ft	m
l_u	laterally unbraced span length	in	mm
M	moment	k-ft	kN-m
M_r	reference design moment	k-ft	kN-m

Symbol	Definition	Imperial	Metric
M'_r	adjusted design moment	k-ft	kN-m
n	number, quantity	unitless	
n	number of fasteners in a row	unitless	
p	length of fastener penetration	in	mm
P	point load, axial compression	k	kN, MN
p_{min}	minimum length of fastener penetration	in	mm
Q	first moment about neutral axis	in³	mm³
q, q_x	area unit load, pressure	lb/ft², k/ft²	N/m², kN/m²
r	radius of a circle or cylinder	in, ft	mm, m
ρ	ratio of steel area to concrete width times depth d	unitless	
R	response modification factor for seismic force	unitless	
$R, R_\#$	reaction	lb or k	N, kN, MN
R_B	slenderness ratio for bending	unitless	
R_r	reference design end reaction	lb	kN
R'_r	adjusted design end reaction	lb	kN
r_x, r_y, r_z	radius of gyration	in	mm
S	snow load	k, k/ft, k/ft², lb, lb/ft, lb/ft²	kN, kN/m, kN/m², N, N/m, N/m²
S	section modulus	in³	mm³
s	spacing	in	mm
T	tension	lb or k	N, kN, MN
t	thickness	in	mm
t_m	main member thickness	in	mm
T_r	reference design shear	lb	kN
T'_r	adjusted design shear	lb	kN
t_s	side member thickness	in	mm
V	shear	lb or k	N, kN, MN
v	unit shear	lb/ft	kN/m

Symbol	Definition	Imperial	Metric
W	reference withdrawal strength of fastener	lb/in pen	kN/mm pen
w	line load, or uniform load	lb/ft	kN/m
W	wind load	k, k/ft, k/ft^2, lb, lb/ft, lb/ft^2	kN, kN/m, kN/m^2, N, N/m, N/m^2
W	weight	lb or k	N, kN, MN
W	reference withdrawal strength of fastener	lb	kN
W'	adjusted withdrawal strength of fastener	lb	kN
W, W'	width of diaphragm	ft	m
w_D	line dead load	lb/ft	kN/m
w_L	line live load	lb/ft	kN/m
w_S	line snow load	lb/ft	kN/m
w_u	factored line load	lb/ft	kN/m
x	geometric axis, distance along axis	unitless	
y	geometric axis, distance along axis	unitless	
Z	reference lateral strength of fastener	lb	kN
z	geometric axis, distance along axis	unitless	
Z'	adjusted lateral strength of fastener	lb	kN
$Z_\perp$	reference lateral design strength of dowel fastener with all wood members loaded perpendicular to grain	lb	kN
$Z_\parallel$	reference lateral design strength of dowel fastener with all wood members loaded parallel to grain	lb	kN
$Z_{m\perp}$	reference lateral design strength of dowel fastener with main member loaded perpendicular to grain	lb	kN
$Z_{s\perp}$	reference lateral design strength of dowel fastener with side member loaded perpendicular to grain	lb	kN

Note: # indicates a general case of symbol and subscript, or subscript and symbol. It can be replaced with a letter or number, depending on how you want to use it. For example $R_\#$ may become R_L for left-side reaction. Similarly, $\#_c$ may become P_c, indicating a compressive point load

Imperial and Metric Conversion Tables

Appendix 9

Table A9.1 Unit conversion table

Imperial to Metric			
	ft	0.305	m
	ft^2	0.093	m^2
	ft^3	0.028	m^3
	in	25.4	mm
	in^2	645.2	mm^2
	in^3	16,387	mm^3
	in^4	416,231	mm^4
	k	4.448	kN
	k/ft	14.59	kN/m
	k/ft^2	47.88	kN/m^2
Multiply	k/ft^3	157.1 (By)	kN/m^3
	k/in^2 (ksi)	6.895	MN/m^2 (MPa) (To get)
	k-ft	1.356	kN-m
	lb, lb$_f$	4.448	N
	lb/ft	14.59	N/m
	lb/ft^2(psf)	47.88	N/m^2 (Pa)
	lb/ft^3	0.157	kN/m^3
	lb/in^2	6,894.8	N/m^2
	lb-ft	1.355	N-m
	lb$_m$	0.454	kg
	mph	1.609	kmh

Metric to Imperial

Multiply	By	To get
m	3.279	ft
m^2	10.75	ft^2
m^3	35.25	ft^3
mm	0.039	in
mm^2	0.0016	in^2
mm^3	6.10237E-05	in^3
mm^4	2.40251E-06	in^4
kN	0.225	k
kN/m	0.069	k/ft
kN/m^2	0.021	k/ft^2
kN/m^3	0.0064	k/ft^3
MN/m^2 (MPa)	0.145	k/in^2 (ksi)
kN-m	0.738	k-ft
N	0.225	lb, lb_f
N/m	0.069	lb/ft
N/m^2 (Pa)	0.021	lb/ft^2(psf)
kN/m^3	6.37	lb/ft^3
N/m^2	1.45E-04	lb/in^2
N-m	0.738	lb-ft
kg	2.205	lb_m
kmh	0.621	mph

Glossary

adjusted design value	reference design value (stress or stiffness) multiplied by applicable adjustment factors
adjustment factor	strength modifier accounting for various influences, such as temperature and size
anisotropic	material properties vary in each direction
apparent stiffness	bending stiffness parameter for I-joists, combining E and I
area load	load applied over an area
area, cross-sectional	area of member when cut perpendicular to its longitudinal axis
area, net	cross-sectional area reamaining after holes or cuts are made
ASD	allowable stress design; factors of safety are applied to the material strength
aspect ratio	ratio of length to width in diaphragms, or height to length in shear walls
axial	action along length (long axis) of member
axis	straight line that a body rotates around, or about which a body is symmetrical
base shear	horizontal shear at base of structure due to lateral wind or seismic forces
beam	horizontal member resisting forces through bending

beam stability factor	material adjustment factor accounting for beam member slenderness (propensity for rolling over)
bearing area factor	material adjustment factor for cross-grain compression away from a member end
bearing wall	wall that carries gravity loads
blocking	small, solid wood member placed between primary members to stabilize and connect them
boundary	line of diaphragm aligning with a shear wall or frame below
box nail	nail with thinner shank and head than common nail
brace	member resisting axial loads (typically diagonal), supports other members
braced frame	structural frame whose lateral resistance comes from diagonal braces
bridging	small bracing members used to brace joists, rafters, and trusses at discrete points along their length
buckling	excess deformation or collapse at loads below the material strength
buckling stiffness factor	material adjustment factor for truss top chords connected by structural panel sheathing
capacity	ability to carry load, related to strength of a member
cellulose	wood fibers giving it axial strength
check	wood crack that runs perpendicular to the grain
chord	truss or diaphragm element resisting tension or compression forces
code	compilation of rules governing the design of buildings and other structures
collector	see drag strut
column	vertical member that primarily carries axial compression load, supports floors and roofs

column stability factor	material adjustment factor accounting for column member slenderness (propensity for buckling)
common nail	typical nail
component	single structural member or element
compression	act of pushing together, shortening
connection	region that joins two or more members (elements)
construction documents	written and graphical documents prepared to describe the location, design, materials, and physical characteristics required to construct the project
contraction joint	groove creating weakened plane in concrete to control crack locations due to dimensional changes
couple, or force couple	parallel and equal, but opposite, forces, separated by a distance
creep	slow, permanent material deformation under sustained load
cross laminated timber	formed by gluing or dovetailing wood boards, in alternating directions
cruck frame	curved timber supporting a roof
curvature factor	material adjustment factor for curved glued laminated timbers
dead load	weight of permanent materials
deflection, δ	movement of a member under load or settlement of a support
demand	internal force due to applied loads
depth	height of bending member, or larger dimension of column
diaphragm	floor or roof slab transmitting forces in its plane to vertical lateral elements
diaphragm factor	material adjustment factor for nailing in a sheathing diaphragm

dimension lumber	lumber that is cut and planed to standard sizes
discontinuity	interruption in material, such as a knot or check
distributed load	line load applied along the length of a member
dowel-type fastener	fastener with round body (nail, bolt, lag screw, drift pin)
drag strut	element that collects diaphragm shear and delivers it to a vertical lateral element
dressed lumber	wood that has been planed to its final size
drift	lateral displacement between adjacent floor levels in a structure
duplex nail	two-headed nail, used in concrete formwork
durability	ability to resist deterioration
eccentricity	offset of force from centerline of a member, or centroid of a fastener group
edge distance	distance between fastener and edge of member
elastic	able to return to original shape after being loaded
element	single structural member or part
empirical design	design methodology based on rules of thumb or past experience
end distance	distance between fastener and end of member
end grain factor	material adjustment factor for dowel-type fastener placed in the end grain
engineered wood	timber products engineered for strength and stiffness, allowing utilization of smaller trees
fixed	boundary condition that does not permit translation or rotation
flat use factor	material adjustment factor for bending members laid flatwise
flexure, flexural	another word for bending behavior
footing	foundation system bearing on soil near the ground surface

force	effect exerted on a body
format conversion factor	material adjustment factor that converts ASD properties to LRFD design
frame	system of beams, columns, and braces, designed to resist vertical and lateral loads
free body diagram	elementary sketch showing forces acting on a body
geometry factor	material adjustment factor for end, edge, and spacing of dowel-type fasteners
girder	beam that supports other beams
glued laminated timber	timber product made from gluing 2× material into larger beams and columns
grading	process of assigning quality rating (related to strength and stiffness) to wood
grain	direction of primary wood fiber
gravity load	weight of an object or structure, directed to the center of the Earth
group action factor	material adjustment factor for a group of bolts (or larger dowel-type fasteners)
header	beam across an opening
I-joist	timber bending member made from sawn or SCL flanges and structural panel webs, formed in the shape of an 'I'
incising factor	material adjustment factor for the effect of preservative notches
indeterminate	problem that cannot be solved using the rules of static equilibrium alone, number of unknowns greater than number of static equilibrium equations
inelastic	behavior that goes past yield, resulting in permanent deformation
isolation joint	separation between adjacent parts of the structure, to allow relative movement and avoid cracking

isotropic	material properties are the same in each direction
jamb studs	bearing studs below a beam or header
king post	column that acts in tension between the apex of a truss and tie beam
king stud	stud that goes past a header to the floor line above
knot	hard portion of wood where a tree branch grew
lag screw	threaded fastener, similar to a bolt, that is screwed into wood from one side
lam	layer of 2× material in glued laminated timber
lateral load	load applied in the horizontal direction, perpendicular to the pull of gravity
lateral torsional buckling	condition where beam rolls over near the middle owing to inadequate bracing for the given load
layup	combination of 2× material in glued laminated timber
lignin	wood material that holds the cellulose (fibers) together
live load	load from occupants or moveable building contents
live load reduction	code-permitted reduction when area supported by a single element is sufficiently large
load	force applied to a structure
load combination	expression combining loads that act together
load duration factor	material adjustment factor accounting for length of time a load is applied, ASD
load factor	factor applied to loads to account for load uncertainty
load path	route a load takes through a structure to reach the ground
long-term deflection	deflection due to sustained loads, such as dead loads

LRFD	load and resistance factor design, also called strength design
LSL	laminated strand lumber, formed by gluing together chips of wood
LVL	laminated veneer lumber, formed by gluing together thick layers of veneer
metal plate connector	proprietary, engineered connector
minimum modulus of elasticity, E_{min}	lower bound of material stiffness parameter, used for stability calculations
modulus of elasticity, E	material stiffness parameter, measure of a material's tendency to deform when stressed
moment arm	distance from a support point at which a force acts
moment frame	structural frame whose lateral resistance comes from rigid beam–column joints
moment, M	twisting force, product of force and the distance to a point of rotation
moment of inertia, I	geometric bending stiffness parameter, property relating area and its distance from the neutral axis
nail	dowel fastener pounded into the wood
NDS	national design specificaion, governing timber design code in the United States
neutral axis	axis at which there is no lengthwise stress or strain, point of maximum shear stress or strain, neutral axis does not change length under load
nominal strength	element strength, typically at ultimate level, prior to safety factor application
oriented strand board	structural panel made from chips of wood, oriented along the panel length and glued together
orthotropic	material with different properties in two or more directions
penetration	depth a nail or lag screw goes into the member away from the head

pin	boundary condition that allows rotation but not translation
plastic	occurs after yield, where material experiences permanent deformation after load is removed
point load	concentrated load applied at a discrete location
point of inflection	point in deflected shape where there is no moment, deflected shape changes direction
preboring	drilling a hole to allow the fastner to be installed or to reduce splitting in driven fasteners
pressure	force per unit area
PSL	parallel strand lumber, formed by gluing strand-like chips of wood together, oriented roughly in the same direction
radius of gyration	relationship between area and moment of inertia used to predict buckling strength
reaction	force resisting applied loads at end of member or bottom of structure
reference design value	basic material strength or stiffness
repetitive member factor	material adjustment factor for potential load sharing between closely spaced and connected members
resistance factor	LRFD material safety factor
resultant	vector equivalent of multiple forces
right-hand rule	positive moment is in direction of thumb when fingers are wrapped in a counter-clockwise direction around the axis of rotation
rigid	support or element having negligible internal deformation
rivet	rectangular-shaped nail used for heavy timber connections
roller	boundary condition that allows rotation, but limits translation in only one direction
row of fasteners	fasteners in a line parallel to the applied force

rupture	complete separation of material
safety factor	factor taking into account material strength or load variability
sawn lumber	wood products made from solid trees and formed by sawing
section modulus	geometric bending strength parameter
Seismic Design Category	classification based on occupancy and earthquake severity
seismic load	force accounting for the dynamic response of a structure or element due to an earthquake
seismic force resisting system	portion of structure designed to resist earthquake effects
shake	wood crack that runs parallel to the grain
shank	shaft portion of a nail
shear	equal, but opposite, forces or stresses acting to separate or cleave a material, like scissors
shear	relative sliding motion in a member, similar to that of scissors
shear plate	high-strength connection made from placing a steel plate into a round groove, thereby engaging a large surface area
shear reduction factor	material adjustment factor reducing shear stresses in glued laminated timbers
shear wall	wall providing lateral resistance for structure
short-term deflection	deflection due to intermittent loads, such as live or snow loads
sign convention	method of assigning positive and negative values to the direction of loads, reactions, and moments
simplifying assumption	assumption that makes the problem easier to solve, but is realistic
sinker nail	nail with tapered head that presses flush into the wood

size factor	material adjustment factor accounting for property changes with size in sawn timber
slender	member that is prone to buckling
snow load	load from fallen or drifted snow
spacing	distance between fasteners
spacing	center-to-center distance between adjacent items
span length	clear distance between supports
special seismic systems	structural systems specifically detailed to absorb seismic energy through yielding
specific gravity	ratio of density of a material divided by the density of water
spike	large nail
split	wood crack that runs neither parallel nor perpendicular to the grain
split ring	high-strength connection made from placing a rolled steel ring into a round groove, thereby engaging a large surface area
stability	structure's resistance to excessive deformation or collapse at loads below the material strength, opposite of buckling
stiffness	resistance to deformation when loaded
strength	material or element resistance to load or stress
strength design	load and resistance factor design; safety factors applied to the loads and materials
stress (f, τ)	force per unit area
stress block	rectangular simplification of the compressive stress in a concrete section
stress interaction factor	material adjustment factor for tapered glued laminated timbers
structural analysis	determination of forces, moments, shears, torsion, reactions, and deformations due to applied loads

structural composite lumber	timber product made from gluing veneers, chips, or smaller components together to form larger, stronger members
structural integrity	ability of structure to redistribute forces to maintain overall stability after localized damage occurs
structural panel	plywood or oriented strand board used to cover floors, roofs, and walls
structural system	series of structural elements (beams, columns, slabs, walls, footings) working together to resist loads
support	either the earth or another element that resists movement of the loaded structure or element
temperature factor	material adjustment factor for sustained temperature
tension	act of pulling apart, lengthening
time effect factor	material adjustment factor accounting for length of time a load is applied, LRFD
toe-nail	nail placed diagonally, from the face of one member into the face of another
toe-nail factor	material adjustment factor for a nail placed in toe-nail fashion
tributary area	area supported by a structural member
tributary width	width supported by a structural member, usually a beam, joist, or girder
truss	structural member comprised of axially loaded members arranged in triangular fashion
truss plate	gage metal plate, punched to create spikes that connect prefabricated trusses together
unbraced length	length between brace point where a member can buckle
volume factor	material adjustment factor accounting for property changes with volume in glued laminated timber

wet service factor	material adjustment factor for moisture content
width	smaller member dimension
wind load	force due to wind
yield	point at which a material has permanent deformation due to applied loads, start of inelastic region of stress–strain curve

Bibliography

Acts and Ordinances of the Interregnum, 1642–1660. Available at: www.british-history.ac.uk/no-series/acts-ordinances-interregnum (accessed November 8, 2016).

AF&PA. *National Design Specification (NDS) for Wood Construction* (Washington, DC: American Forest & Paper Association, 1997).

AF&PA. *National Design Specification (NDS) for Wood Construction Commentary* (Washington, DC: American Forest & Paper Association, 2005), p. 228.

ANSI/AWC. *National Design Specification (NDS) for Wood Construction* (Leesburg, VA: AWC, 2015).

AF&PA. *Commentary on the National Design Specification (NDS) for Wood Construction* (Washington, DC: American Forest & Paper Association, 1997), p. 140.

ASCE. *Minimum Design Loads for Buildings and Other Structures.* ASCE/SEI 7–10 (Reston, VA: American Society of Civil Engineers, 2010).

ASTM. *Standard Specification for Computing Reference Resistance of Wood-Based Materials and Structural Connections for Load and Resistance Design* (West Conshohocken, PA: ASTM International, 2015).

AWC. *ASD/LFRD Manual, National Design Specifications for Wood Construction* (Leesburg, VA: American Wood Council, 2012).

Breyer, D. E., Fridley, K. J., and Cobeen, K. E. *Design of Wood Structures: ASD* (New York: McGraw-Hill).

"Brief history of William Penn." Available at: www.ushistory.org/penn/bio.htm (accessed August 6, 2016).

CEN. *Eurocode 5: Design of Timber Structures* (Brussels: CEN European Committee for Standardization).

CSA Group. *Engineering Design in Wood* (Toronto: CSA Group).

Develin, D. H. "Some Historical Spots in Lower Merion Township," Daughters of the American Revolution, 1906.

Facenda, D. M. "Merion Friends Meetinghouse: Documentation and Site Analysis," Thesis (Master of Science in Historic Preservation, University of Pennsylvania, PA, 2002).

IBC. *International Building Code* (Washington, DC: International Code Council, 2012).

Keast & Hood. *Merion Friends Structural Assessment Report* (Philadelphia, PA: Keast & Hood. 2005).

McMullin, P. W., and Price, J. S. *Introduction to Structures* (New York: Routledge, 2016).

Mitchell, T. "Structural Materials." In *Introduction to Structures*, edited by Paul W. McMullin and Jonathan S. Price (New York: Routledge, 2016).

Parker, H. *Simplified Design of Roof Trusses for Architects and Builders* (New York: John Wiley, 1941).

reTHINK-WOOD. "Tall Wood/Mass Timber." Available at: www.rethinkwood.com/masstimber (accessed May 26, 2015).

Trautwine, J. C. *Civil Engineer's Reference Book* (Sugar Grove, WV: Chapman & Hall, 1937).

U.S. Department of Agriculture, Forest Service. *History of Yard Lumber Size Standards*, by L. W. Smith and L. W. Wood (Washington, DC: U.S. Government Printing Office, 1964).

Weeks, K. D., and Grimmer, A. E. *The Secretary of the Interior's Standards for the Treatment of Historic Properties* (Washington, DC: National Park Service, 1995).

"William Penn." Available at: www.pennsburymanor.org/history/william-penn/ (accessed August 6, 2016).

Yeomans, D. T. "A Preliminary Study of 'English' Roofs in Colonial America," *Association for Preservation Technology Bulletin* 13, no. 4 (1981): 16. Yeomans acknowledges the assistance of Batcheler in discussing the Merion roof.

Index

Taylor & Francis eBooks

Helping you to choose the right eBooks for your Library

Add Routledge titles to your library's digital collection today. Taylor and Francis ebooks contains over 50,000 titles in the Humanities, Social Sciences, Behavioural Sciences, Built Environment and Law.

Choose from a range of subject packages or create your own!

Benefits for you

- » Free MARC records
- » COUNTER-compliant usage statistics
- » Flexible purchase and pricing options
- » All titles DRM-free.

Benefits for your user

- » Off-site, anytime access via Athens or referring URL
- » Print or copy pages or chapters
- » Full content search
- » Bookmark, highlight and annotate text
- » Access to thousands of pages of quality research at the click of a button.

REQUEST YOUR **FREE** INSTITUTIONAL TRIAL TODAY

Free Trials Available
We offer free trials to qualifying academic, corporate and government customers.

eCollections – Choose from over 30 subject eCollections, including:

Archaeology	Language Learning
Architecture	Law
Asian Studies	Literature
Business & Management	Media & Communication
Classical Studies	Middle East Studies
Construction	Music
Creative & Media Arts	Philosophy
Criminology & Criminal Justice	Planning
Economics	Politics
Education	Psychology & Mental Health
Energy	Religion
Engineering	Security
English Language & Linguistics	Social Work
Environment & Sustainability	Sociology
Geography	Sport
Health Studies	Theatre & Performance
History	Tourism, Hospitality & Events

For more information, pricing enquiries or to order a free trial, please contact your local sales team:
www.tandfebooks.com/page/sales

Routledge
Taylor & Francis Group

The home of
Routledge books

www.tandfebooks.com